CONTROL SYSTEMS
MODELING AND ANALYSIS

CONTROL SYSTEMS MODELING AND ANALYSIS

GERARD VOLAND
Northeastern University

PRENTICE-HALL, INC., Englewood Cliffs, New Jersey 07632

Library of Congress Cataloging-in-Publication Data

Voland, Gerard, G. S.
  Control systems modeling and analysis.

  Bibliography: p.
  Includes index.
  1. Control theory. 2. System analysis. I. Title.
QA402.3.V56 1985   629.8'312   85-12263
ISBN 0-13-171984-X

Editorial/production supervision and
    interior design: *David Ershun/Nancy Menges*
Cover design: *20/20 Services, Inc.*
Manufacturing buyer: *Rhett Conklin*

Printed in the United States of America

10 9 8 7 6 5 4 3 2 1

ISBN 0-13-171984-X   01

PRENTICE-HALL INTERNATIONAL (UK) LIMITED, *London*
PRENTICE-HALL OF AUSTRALIA PTY. LIMITED, *Sydney*
PRENTICE-HALL CANADA, INC., *Toronto*
PRENTICE-HALL HISPANOAMERICANA, S.A., *Mexico*
PRENTICE-HALL OF INDIA PRIVATE LIMITED, *New Delhi*
PRENTICE-HALL OF JAPAN, INC., *Tokyo*
PRENTICE-HALL OF SOUTHEAST ASIA PTE. LTD., *Singapore*
EDITORA PRENTICE-HALL DO BRASIL, LTDA., *Rio de Janeiro*
WHITEHALL BOOKS LIMITED, *Wellington, New Zealand*

This modest work is dedicated to my sister and nephews

> *Norma Bell*
> *Mark Bell*
> *Steven Bell*

to my aunt and uncle

> *Thelma Burke*
> *John M. Burke*

and to the Lord who has given me the opportunity to complete it.

> *And if any man think that he knoweth any thing,*
> *he knoweth nothing yet as he ought to know.*

1 Corinthians 8:2

CONTENTS

PREFACE   xi

1 INTRODUCTION TO SYSTEMS ANALYSIS AND DESIGN   1

   1.1   Objectives   1
   1.2   Modeling, Analysis, Design and Control   1
   1.3   Stability and Accuracy   4
   1.4   General Considerations   6
   1.5   Review   8
          Exercises   9

2 DIFFERENTIAL EQUATIONS AND SYSTEMS MODELING   11

   2.1   Objectives   11
   2.2   Differential Equations   12
   2.3   Transient Solution   14
   2.4   Steady-State Solution   20
   2.5   Linearity and the Response Function   24
   2.6   Multiple System Outputs and Simultaneous Differential
          Equations   31
   2.7   Review   37
          Exercises   40

3 LAPLACE TRANSFORMATIONS   46

3.1   Objectives   46
3.2   Transformation Operators   47
3.3   The Laplace Transformation   48
3.4   The Laplace Transform of Derivatives   52
3.5   Inverse Laplace Transformations   55
3.6   A Substitution of Value: The Shift Theorem   57
3.7   Application to Differential Equations   58
3.8   Final-Value and Initial-Value Theorems   67
3.9   Review   68
        Exercises   70

4 NUMERICAL APPROXIMATION METHODS   73

4.1   Objectives   73
4.2   The Need for Numerical Methods   74
4.3   Newton–Raphson Approximation   75
4.4   Method of Successive Approximations   77
4.5   Secant Method and Regula Falsi   78
4.6   Lin's Method for Quadratic Factors   80
4.7   Nonlinear Ordinary Differential Equations:
        Euler's Method and the Runge–Kutta Methods   88
4.8   Review   98
        Exercises   101

5 TRANSFER FUNCTIONS AND BLOCK DIAGRAMS   106

5.1   Objectives   106
5.2   Transfer Functions   106
5.3   Block Diagrams   107
5.4   Block Diagram Algebra and Reduction   110
5.5   Review   113
        Exercises   114

6 SYSTEM ANALYSIS   117

6.1   Objectives   117
6.2   Zeroth-Order and First-Order Systems   118
        6.2.1   Zeroth-Order Systems, 118
        6.2.2   First-Order Systems, 119

6.3   General First-Order Systems   120
        6.3.1   Response to an Applied Step Function, 120
        6.3.2   Response to an Applied Impulse Function, 125
        6.3.3   Response to an Applied Sinusoidal Function, 127
6.4   General Second-Order Systems   129
6.5   System Components and System Analogies   140
6.6   A Management-Production System   156
        6.6.1   Model and Analysis, 156
        6.6.2   Software and Analysis, 158
6.7   Continuous System Simulation   167
6.8   Review   167
        Exercises   168

7 ACCURACY, STEADY-STATE ERROR, AND CONTROL ACTIONS   179

7.1   Objectives   179
7.2   Accuracy and Steady-State Error   180
        7.2.1   Steady-State Error and System Type, 180
        7.2.2   Steady-State Error for a Step Input, 182
        7.2.3   Steady-State Error for a Ramp Input, 184
        7.2.4   Steady-State Error for a Parabolic
                (Acceleration) Input, 186
        7.2.5   Summary, 187
7.3   Control Actions   188
        7.3.1   General Control Action, 188
        7.3.2   Two-Position or On-Off Control, 189
        7.3.3   Proportional Control, 189
        7.3.4   Integral Control, 190
        7.3.5   Proportional-Plus-Integral Control, 191
        7.3.6   Proportional-Plus-Derivative Control, 192
        7.3.7   Proportional-Plus-Derivative-Plus-Integral
                Control, 194
7.4   Application of Various Control Actions   195
        7.4.1   Example Process, 195
        7.4.2   Application of Proportional Control, 196
        7.4.3   Application of Proportional-Plus-Integral
                Control, 197
        7.4.4   Application of Proportional-Plus-Derivative-
                Plus-Integral Control, 198
7.5   Basic Principle for Control Generation   200
7.6   Review   201
        Exercises   203

8 STABILITY   211

8.1   Objectives   211
8.2   Routh–Hurwitz Stability Criterion   211
8.3   Root-Locus Method   220
        8.3.1   Root-Loci Plots, 220
        8.3.2   Root-Loci Procedure, 222
8.4   Review   229
        Exercises   230

9 ALTERNATIVE MODELING AND ANALYSIS SCHEMES   235

9.1   Objectives   235
9.2   Frequency Response   236
9.3   Sampled-Data Systems and the z-Transform   239
9.4   System Compensation   241
9.5   Nonlinearities   242
9.6   Modern Control Theory and State-Space Modeling   242
9.7   Review   244
        Exercises   245

APPENDICES   247

A.   Complex Variables   247
B.   Integration by Parts   248
C.   Method of Partial Fractions   249
D.   "Bond Graphing as a Mode of Technical
       Communication" by G. Voland (originally published in
       IEEE Transactions on Professional Communication,
       Volume PC-25, Number 1, March 1982, pp. 35–37)   251

REFERENCES   255

INDEX   261

PREFACE

This text is designed to provide insight and an enhanced appreciation of control systems modeling and analysis. The reader is assumed to be familiar with calculus, physics, and a computer programming language (FORTRAN is used in this text). *The text is structured for use by those with limited familiarity with the mathematical tools and techniques used in control systems analysis and, indeed, in engineering.*

The perspective of the text is that of "classical" control theory, in which behavior in the time domain and the use of transfer functions are emphasized. It is a focused presentation which develops the reader's ability to interpret the physical significance of mathematical results in systems analysis. A novice in control theory can easily be overwhelmed by the abstract nature of the analytical techniques which are used in advanced work; this text carefully *prepares* the reader for more advanced treatments if he/she chooses to delve further into the subject. The last chapter presents the reader with a perspective of such advanced topics as sampled-data systems, state-space modeling, and modern control theory, which will serve as a smooth transition to texts devoted to these topics.

The pedagogical approach of this text distinguishes it from other systems textbooks. Most treatments of systems theory are so comprehensive that the fundamental concepts are lost among the sheer mass of material. The current work focuses on these fundamental concepts; *learning objectives* are given at the beginning of each chapter in order to guide the reader through the material. These objectives are "performance-oriented"; that is, the reader is expected to be able to *demonstrate* his/her achievement of each objective. There is a *review section* at the end of each chapter so that the reader may obtain a perspective of the topics treated in a given chapter. In addition, there are more than 240 *exercises* in the text with which the reader may sharpen his/her skills. Throughout the text, I have included numerous *solved examples* through which the textual material is related to actual contemporary engineering systems.

Chapter 1 presents a brief historical perspective of control theory and its applications in civilization, together with a review of the objectives and guidelines which should form a vital part of any control engineering effort. The next three chapters present the mathematical foundations of systems analysis: differential equations, Laplace transformations, and numerical approximation techniques. Differential equations form the mathematical models of physical systems, whereas Laplace transformations provide the basis of the transfer function approach in classical control theory and are used to simplify the mathematical difficulties in systems analysis. Chapter 4 reviews several numerical approximation techniques which are valuable in systems analysis (particularly if one uses higher-order mathematical models for systems). Such numerical techniques provide the basis for computer simulation software in systems engineering. This chapter on numerical techniques is an unusual inclusion in a systems textbook; most texts present the (very valuable) root-locus technique for graphically determining the roots of the characteristic equation for a system in the complex (Laplace) *s*-plane, but neglect to consider any other techniques for determining these roots. After studying differential equations and Laplace transformations, students often wonder how one goes about determining the roots of a higher-order equation; however, one must usually delay presentation of the root-locus method until some general systems have been analyzed in terms of stability and accuracy (the root-locus method is presented in Chapter 8 of this text). As a result, there is a period during which the student may question the utility of the differential equation and Laplace transformation models of a physical system. A brief presentation of numerical methods, such as that given in Chapter 4, allows one to emphasize the value of the root-locus method relative to the numerical techniques. Furthermore, such an inclusion introduces the reader to the use of the computer in systems analysis. In addition, many students do not complete a course in numerical techniques or, in fact, are never exposed to numerical techniques during their undergraduate studies. The inclusion of these techniques in a systems course is, therefore, one way in which to introduce such methods to the student in a very appropriate applications environment (systems engineering).

Chapter 5 presents transfer functions and the block diagram representation with which one may specify both system organization and functional relationships between system components. Chapter 6 then applies the mathematical tools developed earlier to analyze various types of systems. Chapters

7 and 8 deal with the two major goals of a control systems effort: accuracy (Chapter 7) and stability (Chapter 8). Control actions, steady-state error, the Routh–Hurwitz stability criterion, and the root-locus method are presented in these two chapters. Finally, Chapter 9 reviews several alternative modeling and analysis schemes, including frequency response, sampled-data systems, the z-transform, system compensation, nonlinearities, modern control theory, and state–state modeling. The appendices review such topics as complex variables, integration by parts, the method of partial fractions, and bond graphing.

This book has been developed through the use of prepublication versions of the textual material in undergraduate and graduate systems courses at Northeastern University. As a result, I firmly believe that the published version is a learning tool of significant value for engineering students. I am grateful to the many students who have provided feedback in my classes during the creation of this text.

I would like to thank several people who have aided immeasurably in the development of this text. First, Dean Harold Lurie of the College of Engineering at Northeastern University and Professor David Freeman, formerly Chairperson of the Industrial Engineering and Information Systems Department at Northeastern University, provided the initial support of this project, for which I am grateful. I also thank Professors Ronald Mourant (current Chairperson of the Department of Industrial Engineering and Information Systems at Northeastern University) and Wilfred Rule for their continued encouragement. In addition, I am grateful to Professors Richard Carter and Stewart Hoover for providing me with the opportunity to teach the undergraduate and graduate courses in control theory which are offered by our department. Mrs. Eleanor Lubin, Director of the Northeastern University Custom Book Program, provided the mechanism through which an earlier test version of this text was used in courses at Northeastern University; I thank her for her support. The efforts of Michael Gunderloy, a student in one of my systems classes who volunteered to proofread the manuscript, are also very much appreciated.

Any textbook author is very dependent on the support he/she receives from the publisher. I am very grateful for the guidance and enthusiastic sponsorship which I have received from Prentice-Hall. A totally professional staff has ensured that the production of this text meets the highest standards. In particular, I thank Mr. Matthew I. Fox, Editor-in-Chief and Assistant Vice President of Prentice-Hall's College Book Division, for his unwavering support.

Dr. David M. Pepper of Hughes Research Laboratories in Malibu, California, suggested specific examples of feedback control applied to research efforts; I am grateful for his interest and suggestions.

I have adopted a stylistic aspect of the excellent book on advanced mathematical methods by Bender and Orszag (1978): the inclusion of quotes taken from the work of Sir Arthur Conan Doyle to illuminate textual material. I, too, am a fan of the Sherlock Holmes stories. I thank Messieurs Bender, Orszag, and (of course) Doyle.

Professor William Crochetiere, Chairperson of the Department of Engineering Design at Tufts University, continues to guide me through the world of systems modeling, analysis, and design with his broad expertise in both theory and applications. I happily take this opportunity to thank him for his guidance and patience.

An author of a textbook—and the book itself—are particularly dependent upon the professional ability and personal enthusiasm of the production editor. I have been most fortunate to work with an editor who has been consistently supportive and totally professional: Mr. David Ershun of Total Concept Associates in Brattleboro, Vermont.

Finally, I thank my wife and colleague, Margaret, for her love, understanding, and assistance in this work (particularly in the development of the computer programs), and my mother, Eleanor, for a lifetime of encouragement and support and for proofreading the galleys for this book.

I would like to conclude with a special request to the reader: Please write to me concerning any errors which may appear in the text. In addition, I will be most grateful for any suggestions which will ensure that the next edition of this text is an even more effective learning tool. All contributions which are adopted will be acknowledged in future editions.

Systems modeling and analysis must be mastered by professional engineers if they are to be truly effective within their chosen discipline. I trust that this text will aid students in the development of their modeling and analytical abilities. In addition, I trust that this text will convince the reader that systems modeling and analysis is indeed fun!

Gerard Voland

Boston, Massachusetts

CONTROL SYSTEMS
MODELING AND ANALYSIS

INTRODUCTION TO SYSTEMS ANALYSIS AND DESIGN

The world is full of obvious things which nobody by any chance ever observes.

Sherlock Holmes, The Hound of the Baskervilles
by Sir Arthur Conan Doyle

1.1 OBJECTIVES

Upon completion of this chapter, the reader should be able to:

- Identify the characteristics of control systems which are desirable.
- Explain the reasons for introducing control in systems designs.
- Distinguish between open-loop and closed-loop systems.
- Explain the use of feedback in control systems.
- Identify examples of systems in which feedback is used in control efforts.
- Define the two major goals of control applications.
- Explain the significance and the effect of large time lags and large sensitivity within a closed-loop feedback control system.
- State the factors that must be considered by the systems analyst/designer as he/she develops a systems model.

1.2 MODELING, ANALYSIS, DESIGN, AND CONTROL

Systems modeling, analysis, and design efforts seek to produce specific responses from the operation of the device, process, or other portion of the universe that forms the "system" under consideration. Control systems can be *manipulated* to produce the desired response or output. *Automatic* control systems do not require human or manual actions for proper control.

Why do we seek automatic control in systems design? The answers are numerous and varied: automatic control systems allow us to produce results (e.g., products, behavior) which are *more consistent* than would be obtained from manually controlled systems. Furthermore, automatic control allows us to free human beings from tasks that are monotonous and dangerous; that is, automatic control systems can increase the *safety* of plant operations and provide the opportunity for people to perform more interesting tasks. The *quality* of the system output is also enhanced through the introduction of automatic control. The operation of automatically controlled systems is generally *faster* and more *accurate* than that of manually controlled systems. We could generate several other reasons to justify the introduction of automatic control in the design of a system.

Why do we seek control in systems design? Consider the system shown in block diagram notation in Figure 1.1. A system or process has been designed, together with a control mechanism or "controller," to produce a desired response. The blocks represent components within the entire design between which are transmitted signals (indicated by lines) in the direction indicated by arrowheads. (More correctly, blocks represent the functional relationships that exist between the signals.) The design shown in Figure 1.1 is known as an *open-loop system*, in which there is no monitoring of the *actual* system response.

Figure 1.1   Open-loop system.

Figure 1.2 presents the general structure of a *closed-loop system*, in which the actual system response $C(t)$ is measured and then compared to the *desired* output. (Open-loop systems actually have no "loop.") The advantage of such an introduction of "feedback"—in which one monitors the performance of a system and uses the measured performance as information for the proper control or manipulation of the system—is that it allows the system to respond to any disturbances that act upon it from the environment. (The environment

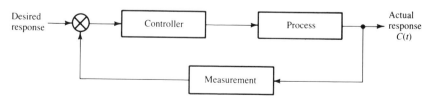

Figure 1.2   Closed-loop system.

is that portion of the universe which is not included in the definition of the system.) An open-loop system behaves properly only if all design specifications are maintained; any disturbance can produce incorrect results. A closed-loop system constantly monitors the output of the process under control; any disturbance that affects the system response will result in an appropriate modification of the system's operation so that the desired output should be obtained.

As an example of a (manually controlled) closed-loop system, consider the case of an automobile being steered by a driver. The driver acts as the monitoring device which determines if the automobile is traveling in the correct direction (e.g., on the highway); he/she also acts as the comparison device which determines if the actual response (direction of travel) is identical to the desired response (desired direction of travel). The driver, together with the steering mechanism of the automobile, also serves as the controller of the system, thereby introducing any changes in the direction of travel so that the desired response is achieved. If the road curves in a certain direction, this information is relayed to the driver through his/her vision and the needed modification in the direction of travel is introduced.

An open-loop system of this type could be achieved if the driver now closes his/her eyes; any disturbances in the operation of the system (e.g., a curve in the road) may now lead to disaster because of the lack of feedback within the system.

Feedback has been extremely important throughout the development of technology (Mayr, 1970). James Watt's centrifugal governor for steam engines (1788) maintained the speed of rotation of the engine by monitoring the actual speed and adjusting the steam inlet valve as needed; any disturbances (e.g., changes in the load or the steam pressure) did not produce incorrect behavior except during short intervals during which the needed adjustment was made. The success of Watt's design provided momentum to the use of feedback in systems design; however, feedback had been used by Ktesibios (in approximately 250 b.c.) in Alexandria in the design of water clocks (in which a float is used to measure the liquid level and thereby control the time-measuring mechanisms). Other float-valve regulators were designed in succeeding generations. A temperature regulator was designed by Cornelis Drebbel (1572–1633) of Alkmaar, Holland, for use in "thermostatic furnaces"; Drebbel's device is recognized as the first feedback system invented in modern Europe that was independent of ancient designs (see Mayr, 1970). Feedback control was also popularly applied to mill designs during the eighteenth and nineteenth centuries.

The interested reader is referred to Mayr's excellent treatment of the early history of feedback control for detailed information about these devices and others designed by creative early engineers (Mayr, 1970).

Bennett (1979) continues to trace the development of control engineering from 1800 through 1930 in his book. *Servomechanisms*, in which the input and output signals are mechanical positions, were developed. Navigational systems, electronic devices, hydraulic components, and so on, were designed to include some form of feedback control. In 1947, Norbert Wiener called the growing field of control and communications theory *cybernetics*, after the Greek word for "steersman" (Wiener, 1948).

Today, we find feedback control in innumerable devices and situations, including the use of robotics in manufacturing processes, thermostats in heating systems, speed control mechanisms in automobiles, automatic pilot devices in airplanes, the vapor pressure control used in gasoline gauges/nozzles, and myoelectrically controlled prosthetic devices. Closed-loop designs are also significant tools in research efforts [for examples of such applications of feedback, see Pepper (1982), Hori et al. (1983), O'Meara (1977), Swigert and Forward (1981), and Forward (1981)].

1.3 STABILITY AND ACCURACY

The two *major* goals of control applications are to produce system behavior which is *stable*, together with a system response which is *accurate* (i.e., identical to the desired system response). These two goals of stability and accuracy are often in conflict, forcing the systems designer to accept a compromise between a response that is accurate and system behavior that is stable.

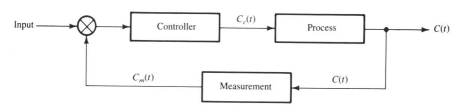

Figure 1.3   Example of a control system.

Instability is the result of *time lags* within the closed loop of the system coupled with relatively large *sensitivity*. (Sensitivity is the amount of *corrective effort per unit error* that is produced within a system in response to a deviation of the actual system output from the desired output.) Figure 1.3 presents a simple closed-loop control system in block diagram form; there is an output signal from each of the components within the loop. Each component requires a finite time interval to generate a response (output signal) to its input signal from the preceding component in the loop. As a result, each component (controller, process, measurement device) produces an *incorrect* output signal—relative to the behavior of the system—during such a time interval. Figure 1.4 illustrates the effect of these time lags or delays on the control of the system output $C(t)$. If the sensitivity of the controller (e.g., the magnitude of its

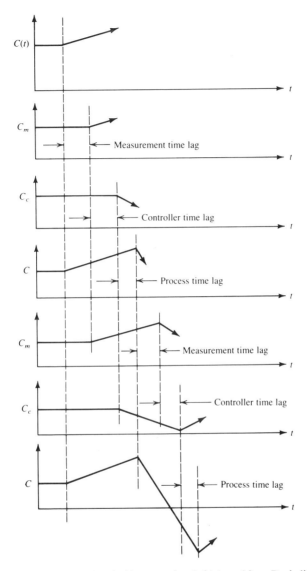

Figure 1.4   Time lags associated with system signals (Adapted from Doebelin, 1962).

output signal) is large *relative* to the time lags within the system, *absolute instability* may occur (Figure 1.5) in which the system response $C(t)$ increases without bound until (for real systems) the system shatters or components are worn beyond use (see Doebelin, 1962).

One attempts to minimize the time lags contained within a system; however, finite delays are inevitable within real systems in which energy (in the form of signals) is transmitted through the loop. As a result, the systems

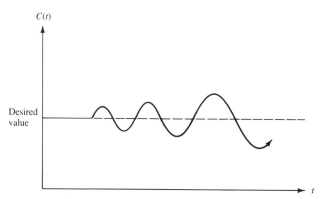

Figure 1.5   Absolute instability reflected in an unbounded system response (Adapted from Doebelin, 1962).

designer must determine the maximum sensitivity that can be introduced into the system without producing a substantial danger of instability. (In addition, the designer should determine the amount of time during which an unstable system may operate before substantial damage or breakdown occurs; an unstable system *may* be acceptable for small operating time intervals. In our treatment, we seek to ensure that a system will *not* behave in an unstable manner.)

The second major goal of accuracy is also dependent on the degree of sensitivity within the system. As sensitivity is increased, the inaccuracy (or error) within the system decreases. However, as sensitivity increases, the danger of instability also increases. As stated earlier, one may be forced to select a compromise between the goals of accuracy and stability. Our treatment seeks to achieve absolute stability, together with maximum (although perhaps not complete) accuracy, in systems design.

1.4 GENERAL CONSIDERATIONS

In addition to the major goals of stability and accuracy, other factors must be considered by the systems designer. These factors may appear to be obvious because they are so fundamental to systems modeling, analysis, design, and simulation; however, one can lose his/her perspective of the problem, the extent of analysis required, and the analytical approach that should be used if he/she fails to consider these additional factors.

Systems modeling (and computer simulation) allows one to experiment with different system designs. The general goal of a modeling effort is to optimize system performance for a given set of realistic operating parameters or for a range of such parameters. The objectives associated with a systems modeling effort include (Voland and Voland, 1983a, 1983b):

- Achieving precise, accurate, and stable system performance
- Ensuring proper corrective response to disturbances that affect system performance
- Determining operating speed
- Maintaining system performance
- Predicting the effect of "auxiliary components" which are inserted into the system to achieve stability, optimize system performance, and so on
- Testing ranges of parameter values
- Investigating different organizational structures for the system
- Minimizing (or eliminating) noise and other undesirable system characteristics

The investigator must also understand the mathematical foundation on which his/her system model is based, together with other factors affecting the ability of the model to predict system behavior accurately. A systems analysis will allow one to solve a set of equations that form the mathematical model of the system; however, one *must* understand the limits of the mathematical model's ability to represent a particular system's behavior; that is, one must be aware of the mathematical (and physical) assumptions and simplifications contained within the model. As a result, the systems analyst should derive the equation(s) for each system component or unit so that a direct one-to-one correspondence between the equations and the system components is achieved; in addition, such derivations allow explicit formulation of the assumptions underlying the system model.

Other factors that should be targeted for consideration by the investigator during the initial system modeling effort are:

- Identification of all significant system components
- Identification of all relevant system parameters and variables
- The range of variation in parameter values that should be investigated
- The structural organization of the system, including the feasibility of alternative organizations
- The desired results (e.g., a description of the system response, maximum accuracy that can be achieved, the dependence of system performance on selected variables)
- Specification of how one will recognize or measure successful completion of the project effort
- The variables that can be controlled and the methods by which one will control these variables in the actual (physical) system

- Specification of the functional dependence on time of the system's input(s)
- Identification of all assumptions contained within the system model, including a justification of the need for such assumptions and evidence of the validity of these assumptions for the system under analysis

In addition, if one plans to simulate the behavior of the system through the use of computer software (computer simulation), several factors associated with the choice of a software package (or the development of an original software package) must be considered:

- The accuracy of simulation that is needed for the investigation
- The availability and the accessibility of existing software packages for system simulation
- The appropriateness of existing simulation software for the particular problem under analysis
- The costs associated with a given simulation effort relative to the expected benefits of such an effort

Based on this review of *availability, accessibility, accuracy, appropriateness,* and *costs,* one may then decide to use existing software or to develop a simple software model of the particular system with which he/she is concerned.

In conclusion, one must not make arbitrary decisions about the system's organization, physical behavior, mathematical model, need for simulation, and so on. Each of these areas must be carefully reviewed by the investigator in terms of the particular objectives sought and the constraints that limit his/her ability to model or simulate the system's behavior.

In the next chapter, we begin to develop the mathematical tools that we will need to model, analyze, and design automatic control systems.

1.5 REVIEW

In summary, we have reviewed the following topics, facts, relationships, or concepts in this chapter.

- Control systems can be manipulated to produce the desired response or output.
- In general, automatically controlled systems produce results which are more consistent in quality; furthermore, these results can be produced faster and more safely—with greater accuracy—than would be achieved through the use of manually controlled systems.

- Open-loop systems do not allow the controlled process to be adjusted in response to a disturbance from the system's environment, whereas closed-loop systems—through the use of feedback—can be modified in response to such a disturbance.
- Feedback has been used in systems design efforts throughout the development of technology—from the water clocks of Ktesibios through the current use of robots in manufacturing and production processes.
- *Stability* and *accuracy* are the two major goals of control efforts.
- Instability within a system is the result of large time lags or delays combined with relatively large sensitivity (corrective effort per unit error).
- Accuracy is also dependent on sensitivity; as sensitivity is increased, error in the actual system response can be decreased. However, as sensitivity increases, the danger of unstable system behavior is also increased.
- Additional objectives and considerations for a systems modeling and analysis effort were identified. In particular, one must understand the limits of a (mathematical) model's ability to represent a particular system's behavior.

EXERCISES

1.1. Identify two examples of open-loop control systems; use block diagram notation to describe the relationships between system components.

1.2. Identify two examples of closed-loop control systems; use block diagram notation to describe the relationships between system components.

1.3. Define the characteristics of closed-loop control systems which are desirable (usually). Define the characteristics of closed-loop control systems which are not desirable (usually).

1.4. Identify an example of a closed-loop feedback system that is:
  (a) Mechanical
  (b) Electrical
  (c) Hydraulic
  (d) Thermal
  (e) Applied to a social or economic process

1.5. List the reasons for introducing automatic control in systems designs.

1.6. List the reasons for introducing manual control in systems designs.

1.7. Explain the difference between open-loop systems and closed-loop systems.

1.8. Describe the circumstances in which one would select open-loop design instead of closed-loop design for a control system.

1.9. Define stability and accuracy; explain why one seeks to achieve these two goals in control applications.

1.10. Explain the effect of large time lags or delays in a closed-loop feedback system. In addition, explain the significance of sensitivity within a control system in terms of stability and accuracy.

1.11. Define the term "absolute instability" as it applies to systems behavior.

1.12. List as many reasons as possible for the systems designer to consider different organizational structures for the same system.

1.13. Explain the importance of identifying the ability of a mathematical system model to predict system behavior accurately. Explain how one identifies such an ability.

1.14. Identify all factors that should be considered during the modeling, analysis, and design of a system. Which of these factors are most important? Does the relative significance of certain factors change in different applications? (If the relative significance does change in different applications, give three examples.)

DIFFERENTIAL EQUATIONS AND SYSTEMS MODELING

Like all other arts, the Science of Deduction and Analysis is one which can only be acquired by long and patient study, nor is life long enough to allow any mortal to attain the highest possible perfection in it. Before turning to those moral and mental aspects of the matter which present the greatest difficulties, let the inquirer begin by mastering more elementary problems.

Sherlock Holmes, A Study in Scarlet *by Sir Arthur Conan Doyle*

2.1 OBJECTIVES

Upon completion of this chapter, the reader should be able to:

- Evaluate differential equations.
- Identify the transient solution of a response function.
- Identify the steady-state solution of a response function.
- Explain the physical significance of the roots of a characteristic equation for a system; in particular, the reader should be able to state the physical significance of:
  (a) Purely real roots
  (b) Purely imaginary roots
  (c) Roots with negative real portions
  (d) Roots with positive real portions
  (e) Roots located at the origin of the complex plane
- Determine the stability of a system through an evaluation of its characteristic equation.
- Use differential operator notation.
- Obtain a linear approximation of a nonlinear function through application of Taylor's series.
- Evaluate simultaneous differential equations via the method of common coefficients or Cramer's rule.

2.2 DIFFERENTIAL EQUATIONS

The behavior of dynamic systems can be described by differential equations. Most physical systems demonstrate linear behavior within small ranges of variable values; their behavior can then be described by linear differential equations.

Differential equations simply allow us to relate the input(s) of the system to the output(s) of the system. The inputs are known as the *forcing functions* [usually denoted as $f(x)$, where x is the independent variable; x will usually represent the time in the physical systems with which we will be concerned]. The system output will be denoted as $y(x)$, representing the dependent variable or *response function* of the system; $y(x)$ is the solution of the differential equation that describes the system's behavior.

In Chapter 3 we will substitute algebraic equations for linear differential equations by using the Laplace transformation, thereby simplifying our analysis of the mathematical description of the system's behavior. Initially, however, we concentrate on the techniques that are available to solve differential equations.

The two categories of differential equations are as follows:

1. *Ordinary* differential equations, in which the dependent variables are functions of only a single independent variable, for example,

$$y = y(t)$$

2. *Partial* differential equations, in which the dependent variables are functions of two or more independent variables, for example,

$$y = y(t, \alpha, \beta, \phi)$$

Solutions to differential equations relate variables without the use of derivatives. Such solutions can be obtained through a variety of methods:

1. Exact analytical methods, which produce exact and general solutions.
2. Stepwise numerical or graphical methods, which produce approximate solutions.
3. Analogy methods, in which one uses the fact that many different types of systems (mechanical, electrical, thermal, economic, etc.) behave according to the same types of general differential equations. As a result, the solutions to these equations are similar in form. One can then use this similarity to identify quickly the solution to an equation that describes the behavior of one type of system (e.g., mechanical) because one is familiar with the solution to an equation that describes an analogous system (e.g., an electrical system).

We will concentrate on linear differential equations with constant coefficients since most of the physical systems with which we will deal can be described (at least for a narrow band of variable values of interest) by such equations. The mathematically necessary conditions for linearity are:

1. *Principle of superposition:* Assume that an excitation (or forcing function) $f_1(x)$, when applied to a system, produces a response $y_1(x)$. An excitation $f_2(x)$ produces a different response $y_2(x)$. If the system behaves according to linearity, the simultaneous application of both excitations will produce a response that is equal to the sum of the individual responses $y_1(x)$ and $y_2(x)$:

$$f_1(x) + f_2(x) \to y_1(x) + y_2(x)$$

2. *Principle of homogeneity:* Assume that an excitation $f_1(x)$, when applied to a linear system, produces a response $y_1(x)$. If one then scales the applied excitation according to some factor k, the response will then also be scaled by this same factor:

$$kf_1(x) \to ky_1(x)$$

Physically, the system is responding to each of the applied excitations; the total response of the system is then the sum of the individual responses to each of the excitations.

Linear differential equations relate the dependent variable and its derivatives in linear combinations. Examples of linear equations with variable coefficients:

$$5x^2 \frac{d^3y}{dx^3} - (1 - \tan x) \frac{d^2y}{dx^2} + 4xy = e^x$$

$$3 \frac{d^2y}{dx^2} + (\sin x) \frac{dy}{dx} - 135 = 0$$

Examples of nonlinear equations:

$$y \frac{d^2y}{dx^2} + \tan y = 15x$$

$$\sin y \frac{d^3y}{dx^3} + 5y^2 \frac{dy}{dx} = 4x^3$$

Again, our focus will be on *linear equations with constant coefficients,* that is, equations of the form

$$a_n \frac{d^ny}{dx^n} + a_{n-1} \frac{d^{n-1}y}{dx^{n-1}} + \cdots + a_1 \frac{dy}{dx} + a_0 y = f(x) \qquad (2.1)$$

where $a_0, a_1, \ldots, a_n$ represent the $(n + 1)$ constant coefficients appearing in the equation. (Please note that some of these coefficients may be equal to zero, that is, certain terms in this general equation may not appear in specific equations.) Examples of such equations:

$$4 \frac{d^4 y}{dx^4} + 9 \frac{dy}{dx} - 5y = 4x + 5e^{2x}$$

$$3 \frac{d^3 y}{dx^3} - 5.8 \frac{d^2 y}{dx^2} + 2y = \sin x$$

The solution $y(x)$ for the general equation (2.1) can be expressed in the form

$$\boxed{y = y_{ts} + y_{ss}} \tag{2.2}$$

where y_{ts} represents the *transient solution*, that portion of the total response function y which is *independent of the forcing function* $f(x)$ applied to the system. This transient solution, y_{ts}, can always be determined. (y_{ts} is also known as the *complementary function*.)

The remaining portion of the response function y is denoted by y_{ss}; it is known as the *steady-state solution* (or the *particular integral*). y_{ss} may be very difficult to determine for certain forcing functions $f(x)$. y_{ss} is that portion of the response function which *is* dependent on $f(x)$.

2.3 TRANSIENT SOLUTION

As noted above, the transient solution y_{ts} is that portion of the response function y that is independent of the applied forcing functions affecting the system. As a result of this independence, y_{ts} is totally dependent and reflective of the system's *design*. The name "transient solution" indicates that we expect y_{ts} to be transient (i.e., not permanent). In fact, we wish to design the system so that y_{ts} *is* transitory. Furthermore, we wish to design the system so that y_{ts} decreases in magnitude with respect to time; eventually, y_{ts} should equal zero so that only the steady-state solution y_{ss} remains as the output y of the system.

In succeeding chapters we develop techniques for ensuring that y_{ts} does not result in an unstable system. At this point of our discussion, however, let us concentrate on the *classical method* for obtaining y_{ts} for equation (2.1).

Following Doebelin (1962), we first introduce *operator notation*. This type of notation will allow us to write linear differential equations of the type (2.1) as algebraic expressions of nth order; once we then determine the n roots of this equation, we can then immediately write an expression for y_{ts}.

The differential operator d/dx is denoted by the symbol D in operator notation. Therefore, the expression (2.1)

$$a_n \frac{d^n y}{dx^n} + a_{n-1} \frac{d^{n-1} y}{dx^{n-1}} + \cdots + a_1 \frac{dy}{dx} + a_0 y = f(x) \qquad (2.1)$$

can be written in operator notation in the form

$$a_n D^n y + a_{n-1} D^{n-1} y + \cdots + a_1 Dy + a_0 y = f(x) \qquad (2.3)$$

where we have used the identity

$$D \equiv \frac{d}{dx} \qquad (2.4)$$

It is imperative that we remember that D is an *operator*; however, we can manipulate D as if it were an algebraic quantity. As a result, we may "factor" the variable y appearing in equation (2.3), thereby obtaining

$$(a_n D^n + a_{n-1} D^{n-1} + \cdots + a_1 D + a_0)y = f(x) \qquad (2.5)$$

which, of course, is equivalent to the expression (2.3).

To determine y_{ts}, we utilize our knowledge that y_{ts} is not dependent on the applied forcing function(s). As a result, we can remove $f(x)$ from equation (2.5) by setting the forcing function equal to zero: that is,

$$(a_n D^n + a_{n-1} D^{n-1} + \cdots + a_1 D + a_0)y = 0 \qquad (2.6)$$

[Again, the differential equation (2.1) describes the behavior of the system under analysis; equation (2.6) will allow us to predict the behavior of this system if it is given an initial amount of energy and then allowed to react without any applied driving force.] If the expression (2.6) is to be valid for any response function y, the expression within parentheses must be equal to zero: that is,

$$a_n D^n + a_{n-1} D^{n-1} + \cdots + a_1 D + a_0 = 0 \qquad (2.7)$$

Equation (2.7) is known as the *characteristic equation* for the system because it specifies the characteristics of the system; in particular, it allows us to determine y_{ts} for the system, which in turn will provide us information about the stability of the system as designed. Again, please note that the characteristic equation (2.7) does not contain any applied forcing function $f(x)$; it describes the system's general, transient behavior which can be expected to appear in addition to any specific, steady-state behavior due to a particular forcing function applied to the system. As a result, *the characteristic equation reflects the general design of the system.*

The characteristic equation (2.7) can be treated as an algebraic expression in D with n roots. Once we have solved for these n roots, we will be able to write the transient solution y_{ts}. Let us denote these n roots by $r_1, r_2, \ldots, r_{n-1}, r_n$. This set of roots may include both real and complex numbers. (See Appendix A for a brief review of complex variables.) Once we have determined the n roots, we may develop an expression for y_{ts} based on the following rules.

Rule 1.   For each root that is *real* and *not repeated* among the n roots, there is a contribution to y_{ts} of the form

$$y_{ts_1} = C_1 e^{r_1 x} \tag{2.8}$$

where r_1 is a real, unrepeated root.

Rule 2.   For each root that is *real* and *repeated m times*, there is a contribution to y_{ts} of the form

$$y_{ts_2} = (C_2 + C_3 x + C_3 x^2 + C_4 x^3 + \cdots + C_{m+1} x^{m-1}) e^{r_2 x} \tag{2.9}$$

where r_2 represents the value of the m roots.

Rule 3.   For each root that is *complex* and *not repeated* among the n roots, there is a contribution to y_{ts} of the form

$$y_{ts_3} = C_{m+2} e^{ax} \sin (bx + \phi) \tag{2.10}$$

where a represents the real portion of the complex root r_3 and b represents the value of the imaginary portion of this root:

$$r_3 = a + jb \tag{2.11}$$

where j denotes the square root of -1: that is,

$$j = (-1)^{1/2} \qquad (2.12)$$

Notice that two constants (C_{m+2} and ϕ) appear in equation (2.10); this is due to the fact that for characteristic equations with *real* constant coefficients a_0, $a_1, a_2, \ldots, a_{n-1}, a_n$ [see equation (2.7)] complex roots must appear in complex conjugate pairs; that is, for root r_3 of the form shown in equation (2.11) there must correspond a root r_4 of the form

$$r_4 = a - jb \qquad (2.13)$$

among the n roots of the characteristic equation. Equation (2.10) gives the contribution to y_{ts} of both r_3 *and* r_4; in other words, equation (2.10) represents the contribution of the entire complex conjugate pair of roots (r_3, r_4) *if the coefficients appearing in the original differential equation (2.1) are purely real numbers. This is very fortunate for our purposes because *these coefficients represent physical parameters* of our system (e.g., specific heat, mass, resistance); as a result, *these coefficients will be purely real numbers in our applications.*

Rule 4.   For each *complex* conjugate pair of roots (r_5, r_6) that are *repeated p times, there is a contribution to y_{ts} of the form

$$\boxed{\begin{aligned} y_{ts_4} &= C_{m+3}\, e^{gx} \sin{(hx + \theta_1)} + C_{m+4}x\, e^{gx} \sin{(hx + \theta_2)} \\ &\quad + \cdots + C_{m+3+p-1}x^{p-1}e^{gx} \sin{(hx + \theta_p)} \end{aligned}} \qquad (2.14)$$

where the complex conjugate roots r_5 and r_6 are equal to

$$r_5, r_6 = g \pm jh \qquad (2.15)$$

that is, g is the real portion of these roots and h denotes the imaginary portion. Notice that y_{ts_4} represents the contribution of $2p$ roots contained within the total set of n roots (i.e., a *pair* of complex conjugate roots which are *repeated p times* among the set of n roots). The appearance of $2p$ constants in equation (2.14) reflects the effect of these $2p$ roots on our analysis (there are p constant coefficients C and p constant phase angles θ in the expression for y_{ts_4}).

Of course, the cases treated by Rules 1 and 3 are simply special cases of those treated by Rules 2 and 4 in which the roots under analysis appear only once among the total set of n roots.

Finally, the complete transient solution y_{ts} is given by the sum of these individual contributions:

$$y_{ts} = y_{ts_1} + y_{ts_2} + y_{ts_3} + y_{ts_4} \qquad (2.16)$$

Example 2.1

As an example of the application of the foregoing four rules, consider a fifth-order differential equation; the five roots of this equation are found to be as follows:

$$r_1 = 1$$
$$r_2 = 3$$
$$r_3 = -2$$
$$r_4 = -5 + j2$$
$$r_5 = -5 - j2$$

y_{ts} is then equal to the sum of the following contributions:

$$y_{ts_1} = C_1 e^{r_1 x}$$
$$\quad\; = C_1 e^x \qquad \text{(due to the real, unrepeated root } r_1)$$
$$y_{ts_2} = C_2 e^{3x} \qquad \text{(due to the real, unrepeated root } r_2)$$
$$y_{ts_3} = C_3 e^{-2x} \qquad \text{(due to the real, unrepeated root } r_3)$$
$$y_{ts_4} = C_4 e^{-5x} \sin(2x + \phi) \quad \text{(due to the unrepeated pair of complex conjugate roots } r_4 \text{ and } r_5)$$

or

$$y_{ts} = y_{ts_1} + y_{ts_2} + y_{ts_3} + y_{ts_4}$$
$$\quad = C_1 e^x + C_2 e^{3x} + C_3 e^{-2x} + C_4 e^{-5x} \sin(2x + \phi)$$

Recall that the independent variable x will usually represent the time t in most of our applications. The transient solution in Example 2.1 would then be written in the form

$$y_{ts} = C_1 e^t + C_2 e^{3t} + C_3 e^{-2t} + C_4 e^{-5t} \sin(2t + \phi) \qquad (2.17)$$

This expression would represent the transient portion of the response function y for the system under investigation—that portion of the system's output which is independent of the particular forcing function applied to the system.

Notice that as time t increases, each of the terms appearing in equation (2.17) either increase or decrease exponentially. As $t \to \infty$, the first two terms approach infinity exponentially, whereas the last two terms exponentially approach zero. Due to the behavior of the first two terms, we conclude that the system is *unstable*; that is, the response y becomes infinite as $t \to \infty$ due to y_{ts}.

What can we conclude, then, about the stability of a system with respect to the values of the roots of the characteristic equation that describes the design of the system? The first two terms in equation (2.17) lead to instability because each has an exponential term raised to a power equal to t multiplied by the value of the real portion of the root. The last term in equation (2.17) could have led to instability if the real portion of the complex roots r_4 and r_5 were positive. In other words, *if the real portion of the root is positive, the result is instability*. (This result is true if the root is purely real or complex.)

Furthermore, if a root is purely real, its contribution to y_{ts} will be a term that either increases exponentially with time t or decreases exponentially with time t (unless the root equals zero, in which case its contribution will be a constant C). If a root is complex, it is part of a complex conjugate pair of roots contained within the set of n roots for the nth-order differential equation that describes the system's behavior. In addition, *the transient solution y_{ts} will behave sinusoidally only if there are complex roots* among this set of n roots. (The sinusoidal behavior will either increase or decrease exponentially, depending on the value of the real portion of the complex root. If the value of the real portion of the complex root is positive, the sinusoidal behavior will increase in amplitude with time t. The sinusoidal behavior will decrease in amplitude if the real portion of the complex root is negative. If the real portion of the root is equal to zero, the sinusoidal behavior will remain constant in amplitude.)

At this point, one can begin to appreciate the amount of information regarding the system's behavior which can be obtained from an analysis of the characteristic equation.

Example 2.2

As another example of the application of our four rules for obtaining y_{ts}, consider the roots of an eighth-order differential equation, which were found to be as follows:

$$r_1 = 5, \quad r_2 = 5, \quad r_3 = 5, \quad r_4 = -3,$$

$$r_5 = -1 + j8, \quad r_6 = -1 - j8, \quad r_7 = -1 + j8, \quad r_8 = -1 - j8$$

The contributions to y_{ts} are then

$y_{ts_1} = (C_1 + C_2 x + C_3 x^2)e^{5x}$ (due to the real, thrice-repeated root value 5)

$y_{ts_2} = C_4 e^{-3x}$ (due to the unrepeated, purely real root r_4)

$y_{ts_3} = C_5 e^{-x} \sin(8x + \phi_1) + C_6 x e^{-x} \sin(8x + \phi_2)$

   (due to the repeated pair of complex conjugate roots)

The complete transient solution is then given by

$$
\begin{aligned}
y_{ts} &= t_{ts_1} + y_{ts_2} + y_{ts_3} \\
&= (C_1 + C_2 x + C_3 x^2)e^{5x} + C_4 e^{-3x} \\
&\quad + C_5 e^{-x} \sin(8x + \phi_1) + C_6 x e^{-x} \sin(8x + \phi_2)
\end{aligned}
\tag{2.18}
$$

(Notice that there are eight constants which will need to be determined: C_1, C_2, C_3, C_4, C_5, C_6, ϕ_1, and ϕ_2—corresponding to the eight roots of the original eighth-order differential equation.)

If the independent variable x were replaced by the time t, is the system of Example 2.2 stable or unstable? The answer is *unstable* because of the behavior of y_{ts_1}; this contribution to y_{ts}, due to the roots r_1, r_2, and r_3, increases exponentially with time (e^{5t}) as a result of the real, positive value $(=5)$ of these three roots.

One might ask how the constant coefficients $(C_1, C_2, \ldots)$ and constant phase angles $(\phi_1, \phi_2, \ldots)$ are determined once the general form of y_{ts} [e.g., that of equation (2.18) in Example 2.2] has been found. To find these constants, one simply applies the known boundary conditions for the system to the response function y and its derivatives appearing in the original differential equation *after* the steady-state solution y_{ss} has been determined. We will review this entire process after we develop a method for determining y_{ss} in the next section.

2.4 STEADY-STATE SOLUTION

The steady-state solution (or particular integral) y_{ss} depends on the applied forcing function(s). As a result, y_{ss} may be very difficult to determine. We will review one method (following Doebelin, 1962) which often leads to a complete specification of y_{ss} for many practical forcing functions. If this method is found to be inappropriate for the system under investigation, another technique will be needed. (We will not review other techniques in this treatment.)

To determine if the method that we will use is appropriate for the forcing function(s) applied to the system under analysis, one must first compute the derivatives of these forcing functions $f(x)$. If

1. after a certain order derivative, all higher-order derivatives are equal to zero, or
2. after a certain order derivative, all higher-order derivatives have the same functional dependence on x as certain lower-order derivatives,

one may conclude that *the method is applicable* and that it will lead to a complete specification of y_{ss}. However, *if*

3. new functions of x appear upon repeated differentiation of $f(x)$,

the method is *not* appropriate for the applied forcing function.
    As examples of these considerations, we present the following cases.

Example 2.3

The applied forcing function $f(x)$ is given by

$$f(x) = 4x^2 \qquad (2.19)$$

Its derivatives are

$$f'(x) \equiv \frac{df}{dx} = 8x \qquad (2.20)$$

$$f''(x) \equiv \frac{d^2f}{dx^2} = 8 \qquad (2.21)$$

$$f'''(x) \equiv \frac{d^3f}{dx^3} = 0 \qquad (2.22)$$

According to condition 1 outlined above, the method that we are about to present for obtaining y_{ss} is applicable for this forcing function.

Example 2.4

The forcing function applied to a system is given by

$$f(x) = \cos 5x \qquad (2.23)$$

The derivatives of $f(x)$ are found to be as follows:

$$f'(x) = -5 \sin 5x \qquad (2.24)$$

$$f''(x) = -25 \cos 5x \qquad (2.25)$$

$$f'''(x) = 125 \sin 5x \qquad (2.26)$$

Since the functional portion of the derivatives repeats itself in higher-order derivatives without new functional forms arising, we may conclude—according to condition 2—that the method we are about to present for determining y_{ss} is applicable for this forcing function.

Example 2.5

The forcing function applied to a system is given by

$$f(x) = 2e^{x^3} \qquad (2.27)$$

The derivatives of this function are as follows:

$$f'(x) = 6x^2 e^{x^3} \tag{2.28}$$

$$f''(x) = 12xe^{x^3} + 18x^4 e^{x^3} \tag{2.29}$$

$$f'''(x) = 12e^{x^3} + 36x^3 e^{x^3} + 72x^3 e^{x^3} + 54x^6 e^{x^3} \tag{2.30}$$

We see that new functional forms appear in succeeding derivatives; condition 3 informs us that the method does not apply to this forcing function. Another method will need to be used in this case.

After we have determined that the method we are about to describe is applicable to the particular case under investigation, we proceed as follows. y_{ss} is postulated to be composed of a sum of terms that consist of the functional portions of the forcing function $f(x)$ and each of its functionally different derivatives, where each term is multiplied by an (as yet) undetermined coefficient. We will denote the functional portion of the forcing function $f(x)$ by $f_v(x)$; that is, $f_v(x)$ is the variable portion of $f(x)$. With this notation, the steady-state solution y_{ss} has the form

$$y_{ss} = Af_v(x) + Bf'_v(x) + Cf''_v(x) + \cdots \tag{2.31}$$

where

$$f'_v(x) = \frac{df_v}{dx} \tag{2.32}$$

$$f''_v(x) = \frac{d^2 f_v}{dx^2} \tag{2.33}$$

and so on, and where $A, B, C, \ldots$ are undetermined coefficients.

Example 2.6

As an application of equation (2.31), consider a system to which is applied the following forcing function:

$$f(x) = 32x^4$$

The steady-state solution y_{ss} is then of the form

$$y_{ss} = Ax^4 + Bx^3 + Cx^2 + Dx + E \tag{2.34}$$

Example 2.7

The forcing function described by

$$f(x) = 10 \cos 5x$$

is applied to a system; the steady-state solution y_{ss} can be expected to be of the form

$$y_{ss} = A \cos 5x + B \sin 5x \tag{2.35}$$

The undetermined coefficients A, B, C, $\therefore$. appearing in equation (2.31) are found by substituting y_{ss} and its derivatives into the original differential equation (2.1) and requiring that the resulting relation be an identity. As an identity, the coefficient of each functionally different term on one side of the equality must be equal to the coefficient of the corresponding functional term on the other side of the equality. An example will clarify this technique.

Example 2.8

Consider the differential equation

$$4\frac{dy}{dx} - 2y = 4x^2 \tag{2.36}$$

[which is an equation of the form

$$a_1\frac{dy}{dx} + a_0 y = f(x)$$

or, in operator notation,

$$(a_1 D + a_0)y = f(x)]$$

The forcing function of equation (2.36) satisfies condition 1, thereby indicating that the steady-state solution y_{ss} can be written in the form of equation (2.31):

$$y_{ss} = Ax^2 + Bx + C \tag{2.37}$$

The first derivative of y appears in equation (2.36); as a result, we must obtain the first derivative of y_{ss} for substitution into this equation:

$$y'_{ss} = 2Ax + B \tag{2.38}$$

Substitution of equations (2.37) and (2.38) into equation (2.36) gives

$$4y'_{ss} - 2y_{ss} = 4x^2$$

or

$$4(2Ax + B) - 2(Ax^2 + Bx + C) = 4x^2 \tag{2.39}$$

We then write all functionally different terms on each side of equation (2.39) and equate coefficients:

$$8Ax + 4B - 2Ax^2 - 2Bx - 2C = 4x^2$$

or

$$-2Ax^2 + (8A - 2B)x + (4B - 2C) = 4x^2 + (0)x + (0) \tag{2.40}$$

Equating coefficients of identical terms on each side of this relation then provides us with the following expressions:

x^2 *term*	$\rightarrow$	$-2A = 4$	(2.41)
x^1 *term*	$\rightarrow$	$8A - 2B = 0$	(2.42)
x^0 *term*	$\rightarrow$	$4B - 2C = 0$	(2.43)

that is, we obtain three equations involving the three unknown coefficients A, B, and C. Solving these three equations simultaneously, we find that

$$A = -2 \tag{2.44}$$

$$B = -8 \tag{2.45}$$

$$C = -16 \tag{2.46}$$

or, upon substitution of these values for the three coefficients into equation (2.37),

$$y_{ss} = -2x^2 - 8x - 16 \tag{2.47}$$

Finally, a *test* of our result [and, in particular, of the algebraic manipulations that led to equations (2.44) through (2.46)] should be performed as follows:

$$y'_{ss} = -4x - 8 \tag{2.48}$$

Substitution of equations (2.47) and (2.48) into the original differential equation (2.36) gives

$$4y'_{ss} - 2y_{ss} = 4(-4x - 8) - 2(-2x^2 - 8x - 16)$$

or

$$4y'_{ss} - 2y_{ss} = 4x^2 + (-16 + 16)x + (-32 + 32)$$

$$= 4x^2$$

which successfully concludes our test of equation (2.47).

   In the next section we treat a system to which multiple forcing functions are applied.

2.5 LINEARITY AND THE RESPONSE FUNCTION

Recall that linear equations must obey the principle of superposition. As a result, the application of a forcing function of the form

$$\boxed{f(x) = f_1(x) + f_2(x) + \cdots + f_n(x)} \tag{2.49}$$

(i.e., the sum of n functionally different forcing functions which are simultaneously applied to the system under investigation) results in a total response function equal to the sum of the n individual responses that would be produced by the n applied forces:

$$\boxed{y_{ss} = y_{ss_1} + y_{ss_2} + \cdots + y_{ss_n}} \tag{2.50}$$

Equation (2.50) states that the system responds to each of the applied forcing functions and that its total response is simply the sum of these individual responses.

Please note that equation (2.50) contains only *steady-state* responses, not transient solutions, because only steady-state solutions are dependent on (and result from) applied forcing functions. The full response function y is equal to the sum of the transient solution y_{ts} and the total steady-state solution y_{ss}, as stated in equation (2.2).

One might ask at this point in our presentation if physical systems usually exhibit linear behavior. Many systems do behave linearly—at least *within specific ranges* of the independent variable(s). Other systems exhibit nonlinear behavior which must be described by nonlinear differential equations; such equations may be *very* difficult to solve analytically. For such nonlinear systems, it is often advantageous to use a linear approximation of the true differential equation that describes system behavior. Linear approximations can be obtained if the nonlinear relationship is described by a smooth function.

We will briefly review a technique for linearization of a nonlinear function. If one restricts the independent variable to a narrow band of values, the dependent function is approximately equal to its tangent line (see Figure 2.1). The equation of the tangent line is simply that of a straight line:

$$y = mx + b \qquad (2.51)$$

where m represents the slope of the straight line and b is its intercept: that is,

$$m = \frac{\Delta y}{\Delta x} \qquad (2.52)$$

$$b = y(0) \qquad (2.53)$$

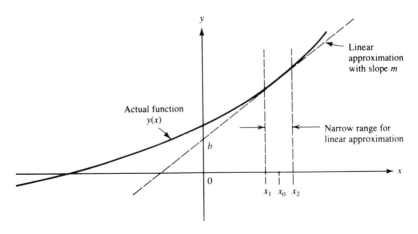

Figure 2.1   Approximating a function $y(x)$ by a linear model.

To obtain a linear approximation of the function $y(x)$, we will use *Taylor's series*. Taylor's series allows us to express a function $y(x)$ in the form of a power series expansion:

$$y(x) = y(x_0) + \left.\frac{dy}{dx}\right|_{x=x_0} \frac{x - x_0}{1!} + \left.\frac{d^2y}{dx^2}\right|_{x=x_0} \frac{(x - x_0)^2}{2!}$$
$$+ \cdots + \left.\frac{d^n y}{dx^n}\right|_{x=x_0} \frac{(x - x_0)^n}{n!} \tag{2.54}$$

[Taylor's series is a *very useful tool* for simplifying functions that need to be analyzed. The inclusion of more terms in the expansion (2.54) increases the accuracy of the approximation. Taylor's theorem, upon which the series expansion is based, was introduced by the mathematician Brook Taylor (1685–1731) in 1715 (see Marion, 1973).] x_0 is the value of the independent variable x about which the expansion is performed—that is, x_0 is the point of tangency on the curve about which the approximation is obtained.

The *linear approximation* is obtained from the Taylor series by retaining only the first two terms of the expansion:

$$y(x) \simeq y(x_0) + \left.\frac{dy}{dx}\right|_{x=x_0} (x - x_0) \tag{2.55}$$

Upon comparison of equations (2.51) and (2.55), the following identifications can be made:

$$b = y(x_0) - \left.\frac{dy}{dx}\right|_{x=x_0} x_0 = \text{intercept} \tag{2.56}$$

and

$$m = \left.\frac{dy}{dx}\right|_{x=x_0} = \text{slope} \tag{2.57}$$

(see Figure 2.1).

Finally, if the dependent variable is a function of *multiple independent variables* (e.g., x, t, θ, λ), we may extend Taylor's series and the linear approximation by expanding about a point defined by x_0, t_0, θ_0, λ_0, and so on, using partial derivatives:

$$y \simeq y(x_0, t_0, \theta_0, \lambda_0) + \frac{\partial y}{\partial x}(x - x_0)$$

$$+ \frac{\partial y}{\partial t}(t - t_0) + \frac{\partial y}{\partial \theta}(\theta - \theta_0) + \frac{\partial y}{\partial \lambda}(\lambda - \lambda_0) + \cdots \tag{2.58}$$

where each of the partial derivatives is evaluated at the point of interest.

As a result of Taylor's series, we may now evaluate nonlinear functions within narrow bands of interest; that is, we may apply the methods discussed earlier for determining y_{ts} and y_{ss} after we obtain a linear approximation of the system's equation(s) through the use of equation (2.55) or (2.58).

We now apply our mathematical tools to analyze a system to which multiple forcing functions are applied.

Example 2.9

Consider a system to which is applied two forcing functions:

$$f_1(x) = 5x \tag{2.59}$$

$$f_2(x) = 2 \sin 4x \tag{2.60}$$

The behavior of this system is described by the differential equation

$$2 \frac{d^2 y}{dx^2} + 3 \frac{dy}{dx} + 2y = 5x + 2 \sin 4x \tag{2.61}$$

[i.e., $f_1(x)$ and $f_2(x)$ are applied simultaneously]. Our task is to determine the full response or output y of this system.

To be more realistic in our notation and our model of the system, let us identify the independent variable x as the time t during which the system operates; in this case, equation (2.61) has the form

$$2 \frac{d^2 y}{dt^2} + 3 \frac{dy}{dt} + 2y = 5t + 2 \sin 4t \tag{2.62}$$

The forcing functions then consist of one that increases linearly with time and one that oscillates with a frequency $\omega = 4$.

In addition, let us introduce the following initial conditions at $t = 0$:

$$y = 0 \tag{2.63}$$

$$y' = 0 \tag{2.64}$$

[A second-order differential equation, such as expression (2.62), requires that two initial conditions be known in order to allow us to find the two constants that will appear in the transient solution y_{ts}. Similarly, a third-order equation will require that three initial conditions be known, a fourth-order equation requires knowledge of four boundary conditions, and so on.]

We are now prepared to determine the response function y.

Step 1: Obtain the transient solution y_{ts}.

Setting the applied forcing functions equal to zero, we obtain the characteristic equation for the system:

$$2\frac{d^2y}{dx^2} + 3\frac{dy}{dt} + 2y = 0 \tag{2.65}$$

or, in operator notation,

$$(2D^2 + 3D + 2)y = 0 \tag{2.66}$$

Of course, the expression in parentheses must be equal to zero if equation (2.66) is valid for arbitrary values of the variable y:

$$(2D^2 + 3D + 2) = 0 \tag{2.67}$$

which is an algebraic second-order expression with root values given by the quadratic equation:

$$r_1, r_2 = \frac{-3 \pm [(3)^2 - 4(2)(2)]^{1/2}}{2(2)} \tag{2.68}$$

$$= -0.75 \pm j\frac{(7)^{1/2}}{4} \tag{2.69}$$

Since r_1 and r_2 form a pair of complex, unrepeated roots, y_{ts} must be of the form given by equation (2.10):

$$y_{ts} = Ce^{-0.75t}\sin\left[\frac{(7)^{1/2}}{4}t + \phi\right] \tag{2.70}$$

The coefficient C and the phase factor ϕ must be determined by using the two known initial conditions after the full response function has been found.

Step 2: Obtain the steady-state solution y_{ss}.

y_{ss} is composed of *two* terms: y_{ss_1}, which is the result of the forcing function $f_1(t)$, and y_{ss_2}, which is the result of $f_2(t)$. Each of the two forcing functions satisfies the necessary conditions (1 and 2) for the method discussed in Section 2.4 (for finding y_{ss}) to be applicable. Thus we may postulate that

$$y_{ss_1} = At + B \tag{2.71a}$$

$$y_{ss_2} = E\sin 4t + F\cos 4t \tag{2.71b}$$

We will also need y'_{ss_1}, y'_{ss_2}, y''_{ss_1}, and y''_{ss_2} for substitution into the original differential equation (2.62).

$$y'_{ss_1} = A \tag{2.72}$$

$$y'_{ss_2} = 4E\cos 4t - 4F\sin 4t \tag{2.73}$$

$$y''_{ss_1} = 0 \tag{2.74}$$

$$y''_{ss_2} = -16E\sin 4t - 16F\cos 4t \tag{2.75}$$

$$= -16y_{ss_2}$$

We next write equation (2.62) for *each* steady-state response (y_{ss_1} and y_{ss_2}), analyzing the effect of *each* forcing function. Therefore,

$$2y''_{ss_1} + 3y'_{ss_1} + 2y_{ss_1} = 5t \tag{2.76}$$

or

$$2(0) + 3(A) + 2(At + B) = 5t \tag{2.77}$$

that is,

$$2At + (3A + 2B) = 5t \tag{2.78}$$

Equating coefficients of similar terms, we find that

$$2A = 5 \quad \rightarrow \quad A = 2.5 \tag{2.79}$$

$$3A + 2B = 0 \quad \rightarrow \quad B = -3.75 \tag{2.80}$$

As a result,

$$y_{ss_1} = 2.5t - 3.75 \tag{2.81}$$

For y_{ss_2}, we have

$$2y''_{ss_2} + 3y'_{ss_2} + 2y_{ss_2} = 2 \sin 4t \tag{2.82}$$

Using equations (2.71), (2.73), and (2.75), this relationship becomes

$$2(-16E \sin 4t - 16E \cos 4t) + 3(4E \cos 4t - 4F \sin 4t) + 2(E \sin 4t + F \cos 4t)$$
$$= 2 \sin 4t \tag{2.83}$$

or

$$(-12F - 30E) \sin 4t + (12E - 30F) \cos 4t = 2 \sin 4t + (0) \cos 4t \tag{2.84}$$

Equating coefficients then gives

$$-12F - 30E = 2 \tag{2.85}$$

$$-30F + 12E = 0 \tag{2.86}$$

Thus we find that

$$E = -\tfrac{5}{87} \tag{2.87}$$

$$F = -\tfrac{2}{87} \tag{2.88}$$

Therefore,

$$y_{ss_2} = -\tfrac{5}{87} \sin 4t - \tfrac{2}{87} \cos 4t \tag{2.89}$$

The total steady-state solution is then given by

$$y_{ss} = y_{ss_1} + y_{ss_2}$$
$$= 2.5t - 3.75 - \tfrac{5}{87} \sin 4t - \tfrac{2}{87} \cos 4t \tag{2.90}$$

Step 3: State the full response function y.

$$y = y_{ts} + y_{ss}$$

$$= Ce^{-0.75t} \sin\left[\frac{(7)^{1/2}}{4}t + \phi\right] + 2.5t - 3.75 - \tfrac{5}{87}\sin 4t - \tfrac{2}{87}\cos 4t \tag{2.91}$$

Step 4: Use the initial boundary conditions to determine the values of the constants appearing in the transient solution y_{ts}.

To apply the initial conditions given by equations (2.63) and (2.64), we needed to find the full response function y; the initial conditions refer to y—not y_{ts} or y_{ss}—so that y must be given, as in equation (2.91). In addition, y' must be determined. Both y and y' are then evaluated at $t = 0$.

$$y' = \frac{dy}{dt}$$

$$= -0.75Ce^{-0.75t}\sin\left[\frac{(7)^{1/2}}{4}t + \phi\right] + \frac{(7)^{1/2}}{4}Ce^{-0.75t}\cos\left[\frac{(7)^{1/2}}{4}t + \phi\right] \tag{2.92}$$

$$+2.5 - \tfrac{20}{87}\cos 4t + \tfrac{8}{87}\sin 4t$$

Finally, the initial conditions (at $t = 0$) are introduced; equations (2.91) and (2.92) then become

$$y(t = 0) = 0$$

$$= C\sin\phi - (3.75 + \tfrac{2}{87}) \tag{2.93}$$

$$y'(t = 0) = 0$$

$$= -0.75C\sin\phi + \frac{(7)^{1/2}}{4}C\cos\phi + 2.5 - \tfrac{20}{87} \tag{2.94}$$

Equation (2.93) states that

$$C\sin\phi = 3.75 + \tfrac{2}{87}$$

$$\simeq 3.773 \tag{2.95}$$

Equation (2.94), together with the result of equation (2.95), then gives

$$C\cos\phi = \frac{4}{(7)^{1/2}}(0.56)$$

$$\simeq 4.88 \tag{2.96}$$

Equations (2.95) and (2.96) allow us to determine the phase angle ϕ:

$$\frac{C\sin\phi}{C\cos\phi} = \frac{3.773}{4.88} = 0.773$$

$$= \tan\phi \tag{2.97}$$

$$\rightarrow \quad \phi = \arctan 0.773 = \tan^{-1} 0.773$$

$$= 37.7 \text{ degrees} \tag{2.98}$$

The value of the coefficient C can then be found from either equation (2.95) or equation (2.96):

$$C = \frac{3.773}{\sin \phi}$$

$$= 6.17 \tag{2.99}$$

The full response function of this system is then

$$y = y_{ts} + y_{ss}$$

$$= 6.17 e^{-0.75t} \sin\left[\frac{(7)^{1/2}}{4} t + 37.7°\right] + 2.5t - 3.75 - \tfrac{5}{87}\sin 4t - \tfrac{2}{87}\cos 4t \tag{2.100}$$

Notice that the first term in expression (2.100) decreases exponentially with time; this term is y_{ts} and the exponential "dampening" of this term is a reflection of the design of the system. Mathematically, this aspect of the system design is contained in equation (2.69), where we see that the real portion of the roots r_1 and r_2 is negative—hence the exponential decrease in y_{ts} with increasing time t. The system is *stable* in terms of its *design*.

However, the system output is *not* bounded, according to equation (2.100). Notice that the final two terms of equation (2.100) indicate that a portion of the full response function y will oscillate with a frequency equal to 4; this is due to the applied forcing function $f_2(t)$, which also oscillates with this frequency. There is also a term $(2.5t)$ which increases linearly with time; this term is due to the other applied input $f_1(t)$ to the system. Since $f_1(t)$ increases *without limit*, the output from the system must also increase without limit! (Of course, we know that an actual system cannot produce an unlimited response and that, instead, the system will simply collapse after reaching a critical point.) The application of $f_1(t)$ is the reason that the system response y is not bounded.

We have restricted our analysis to those systems that produce only one response y; what about systems with *multiple outputs*? We consider such systems in the next section.

2.6 MULTIPLE SYSTEM OUTPUTS AND SIMULTANEOUS DIFFERENTIAL EQUATIONS

We have restricted our attention to systems with only a single response or output y. We now extend our treatment to include systems with multiple outputs.

Systems with n multiple outputs $y_1, y_2, \ldots, y_n$ behave according to the description specified in n multiple, simultaneous differential equations. Operator notation will allow us to treat these equations as n simultaneous algebraic expressions. Two approaches for developing a set of solutions to these equations will be presented.

Approach 1: Multiplication by common coefficients. Consider a set of two differential equations which describe the behavior of a system with two distinct outputs y_1 and y_2 (e.g., a mechanical system with two coupled masses); these equations are

$$a_2 \frac{d^2 y_1}{dt^2} + a_1 \frac{dy_2}{dt} + a_0 y_1 = f_1(t) \qquad (2.101)$$

$$b_3 \frac{d^2 y_2}{dt^2} + b_2 \frac{dy_1}{dt} + b_1 \frac{dy_2}{dt} + b_0 y_2 = f_2(t) \qquad (2.102)$$

Both equations are second-order; each equation includes the two dependent variables y_1 and y_2, necessitating that we solve these equations simultaneously. The coefficients a_2, a_1, a_0, b_3, b_2, b_1, and b_0 represent system parameters, such as mass values and spring constants; as a result, these coefficients must be purely real numbers. These equations may be written in operator notation as

$$(a_2 D^2 + a_0)y_1 + (a_1 D)y_2 = f_1(t) \qquad (2.103)$$

$$(b_2 D)y_1 + (b_3 D^2 + b_1 D + b_0)y_2 = f_2(t) \qquad (2.104)$$

To solve for one of the dependent variables (e.g., y_1), we need to eliminate the other variable (y_2) from equations (2.103) and (2.104). To accomplish this feat, we simply multiply each equation by the coefficient of the variable to be eliminated in the other equation. For example, we would multiply equation (2.103) by the coefficient $(b_3 D^2 + b_1 D + b_0)$ of y_2 in equation (2.104), while simultaneously multiplying equation (2.104) by the coefficient $(a_1 D)$ of y_2 in equation (2.103), thereby obtaining

$$(a_2 D^2 + a_0)(b_3 D^2 + b_1 D + b_0)y_1 + (a_1 D)(b_3 D^2 + b_1 D + b_0)y_2$$
$$= (b_3 D^2 + b_1 D + b_0)f_1(t) \qquad (2.105)$$

$$(b_2 D)(a_1 D)y_1 + (a_1 D)(b_3 D^2 + b_1 + b_0)y_2 = (a_1 D)f_2(t) \qquad (2.106)$$

Notice that the coefficients of y_2 in equations (2.105) and (2.106) are identical (i.e., they are "common coefficients"). If we now subtract equation (2.106) from equation (2.105), the y_2 term is eliminated (as we desired):

$$[(a_2 D^2 + a_0)(b_3 D^2 + b_1 D + b_0) - (b_2 D)(a_1 D)]y_1$$
$$= (b_3 D^2 + b_1 D + b_0)f_1(t) - (a_1 D)f_2(t) \qquad (2.107)$$

Equation (2.107) can be converted to standard differential notation, producing

$$\left[a_2 b_3 \frac{d^4}{dt^4} + a_2 b_1 \frac{d^3}{dt^3} + (a_0 b_3 - a_1 b_2 + a_2 b_0) \frac{d^2}{dt^2} + a_0 b_1 \frac{d}{dt} + a_0 b_0 \right] y_1$$
$$= \left(b_3 \frac{d^2}{dt^2} + b_1 \frac{d}{dt} + b_0 \right) f_1(t) - \left(a_1 \frac{d}{dt} \right) f_2(t) \qquad (2.108)$$

Equation (2.108) is simply a fourth-order differential equation that can be solved for y_1 using the methods discussed earlier in the chapter. A similar equation can be obtained for y_2. A specific example will complete our presentation of this approach.

Example 2.10

Consider the following two simultaneous differential equations:

$$\left(3\frac{d^2}{dt^2} + 5\frac{d}{dt} + 2\right)y_1 + \left(12\frac{d}{dt} + 3\right)y_2 = 4e^{-3t} \tag{2.109}$$

$$\left(4\frac{d}{dt} - 2\right)y_1 + \left(4\frac{d^3}{dt^3} + 2\frac{d}{dt} - 5\right)y_2 = 3\sin 2t \tag{2.110}$$

The independent variable y_2 can be eliminated by use of approach 1; multiplying each of the equations above by the coefficient (in parentheses, including differential operators) of y_2 in the other equation and then subtracting one of the resulting expressions from the other gives

$$\left[\left(3\frac{d^2}{dt^2} + 5\frac{d}{dt} + 2\right)\left(4\frac{d^3}{dt^3} + 2\frac{d}{dt} - 5\right) - \left(4\frac{d}{dt} - 2\right)\left(12\frac{d}{dt} + 3\right)\right]y_1$$

$$= \left(4\frac{d^3}{dt^3} + 2\frac{d}{dt} - 5\right)4e^{-3t} - \left(12\frac{d}{dt} + 5\right)3\sin 2t \tag{2.111}$$

This single equation, which contains only y_1 as a dependent variable, can be written with greater clarity by performing the multiplications of terms in parentheses indicated in this expression and by performing the differential operations on the given forcing functions. The result is as follows:

$$\left(12\frac{d^5}{dt^5} + 20\frac{d^4}{dt^4} + 14\frac{d^3}{dt^3} - 53\frac{d^2}{dt^2} - 9\frac{d}{dt} - 4\right)y_1$$

$$= -476e^{-3t} - 72\cos 2t - 15\sin 2t \tag{2.112}$$

a fifth-order equation that can be solved for y_1.

The foregoing approach (known as the *method of common coefficients*) is useful for those cases in which only two or three simultaneous equations must be solved. The behavior of many systems requires three or more simultaneous equations for its description because of multiple system outputs. In such cases, one often attempts to simplify the analysis of the system by eliminating some of the system responses from the group of outputs that are to be predicted or controlled; that is, certain outputs are intentionally neglected by the analyst because they are insignificant in the total investigation. Of course, such an approach of "intentional neglect" is not very appealing unless it produces an appropriate and accurate result in an efficient manner.

For those cases in which several simultaneous equations *must* be considered in the analysis of a system, one may use determinants in the following technique.

Approach 2: Analysis with determinants (Cramer's rule). Consider a set of simultaneous equations of the form

$$E_{11}y_1 + E_{12}y_2 + E_{13}y_3 + \cdots + E_{1n}y_n = f_1(t)$$

$$E_{21}y_1 + E_{22}y_2 + E_{23}y_3 + \cdots + E_{2n}y_n = f_2(t)$$

$$E_{31}y_1 + E_{32}y_2 + E_{33}y_3 + \cdots + E_{3n}y_n = f_3(t) \qquad (2.113)$$

$$\vdots \qquad \vdots \qquad \vdots \qquad \vdots \qquad \vdots$$

$$E_{n1}y_1 + E_{n2}y_2 + E_{n3}y_3 + \cdots + E_{nn}y_n = f_n(t)$$

that is, n equations with n unknown variables $y_1, y_2, \ldots, y_n$. The coefficients E_{ij} may be equal to zero in some cases; these coefficients may also include differential operators. We may rewrite these equations in a more compact form by introducing determinants. The result is as follows:

$$\begin{vmatrix} E_{11} & E_{12} & E_{13} & \cdots & E_{1n} \\ E_{21} & E_{22} & E_{23} & \cdots & E_{2n} \\ E_{31} & E_{32} & E_{33} & \cdots & E_{3n} \\ \vdots & \vdots & \vdots & \cdots & \vdots \\ E_{n1} & E_{n2} & E_{n3} & \cdots & E_{nn} \end{vmatrix} \begin{vmatrix} y_1 \\ y_2 \\ y_3 \\ \vdots \\ y_n \end{vmatrix} = \begin{vmatrix} f_1(t) \\ f_2(t) \\ f_3(t) \\ \vdots \\ f_n(t) \end{vmatrix} \qquad (2.114)$$

E_{ij} is the element in the ith row and jth column of the "coefficient determinant" which multiplies the variable y_j in the ith equation. With this notation, we may then focus on any of the n dependent variables and obtain a single equation in which only that particular variable appears, together with a series of coefficients and known functions. As a result, we may choose to determine only one of the n variables, the entire set $(y_1, y_2, \ldots, y_n)$ or any combination of variables that we deem to be appropriate. An expression for variable y_3 is given by

$$y_3 = \frac{\begin{vmatrix} E_{11} & E_{12} & f_1(t) & \cdots & E_{1n} \\ E_{21} & E_{22} & f_2(t) & \cdots & E_{2n} \\ E_{31} & E_{32} & f_3(t) & \cdots & E_{3n} \\ \vdots & \vdots & \vdots & \cdots & \vdots \\ E_{n1} & E_{n2} & f_n(t) & \cdots & E_{nn} \end{vmatrix}}{\{E\}} \qquad (2.115)$$

where $\{E\}$ refers to the coefficient determinant given in equations (2.114):

$$\{E\} \equiv \begin{vmatrix} E_{11} & E_{12} & E_{13} & \cdots & E_{1n} \\ E_{21} & E_{22} & E_{23} & \cdots & E_{2n} \\ E_{31} & E_{32} & E_{33} & \cdots & E_{3n} \\ \vdots & \vdots & \vdots & \cdots & \vdots \\ E_{n1} & E_{n2} & E_{n3} & \cdots & E_{nn} \end{vmatrix} \qquad (2.116)$$

Notice that the numerator appearing on the right side of equation (2.115) is a determinant that is identical to $\{E\}$ with the exception that the third column of $\{E\}$ has been replaced with the n forcing functions appearing in the original equations (2.114).

In general, the method of determinants requires that one replace the jth column of the coefficient determinant $\{E\}$ with the n forcing functions given in the original n equations; the resulting determinant is then used as the numerator of the ratio wherein the denominator is $\{E\}$. This quotient is equal to the jth dependent variable y_j. (The jth column of coefficients in $\{E\}$ are the coefficients of y_j in the original simultaneous equations.)

One note of caution: The elements of the determinants appearing in equation (2.115) or, in general, in the equation one obtains for y_j, often include differential operators. These operators may be applied to the known forcing functions appearing in the numerator of the equation, but the operators contained in the denominator $\{E\}$ must be applied to the unknown variable y_j which is sought. The operators contained in $\{E\}$ do *not* act on the other elements E_{ij} of $\{E\}$, only on y_j. The differential equation equivalent to equation (2.115), for example, must then be solved by the techniques discussed earlier in the chapter.

A specific example will clarify the use of this technique.

Example 2.11

Consider the following three simultaneous equations, which include three unknown response functions y_1, y_2, and y_3.

$$2\frac{d^2 y_1}{dt^2} + 4\frac{dy_2}{dt} + 3y_3 = te^{-t} \tag{2.117}$$

$$\frac{dy_1}{dt} + y_2 - 3\frac{dy_3}{dt} = 4t^2 \tag{2.118}$$

$$3\frac{dy_1}{dt} - 2y_1 - 5y_2 - 2y_3 = 3\cos 2t \tag{2.119}$$

Introducing operator notation for the differential terms, we may convert these equations into determinant notation; the result is

$$\begin{vmatrix} 2D^2 & 4D & 3 \\ D & 1 & -3D \\ 3D-2 & -5 & -2 \end{vmatrix} \begin{vmatrix} y_1 \\ y_2 \\ y_3 \end{vmatrix} = \begin{vmatrix} te^{-t} \\ 4t^2 \\ 3\cos 2t \end{vmatrix} \tag{2.120}$$

Let us assume that the particular system response y_1 is needed by the analyst. The method of determinants states that an equation in which y_1 appears as the only unknown variable can be obtained by replacing the first column of elements in the coefficient determinant $\{E\}$ with the known forcing functions appearing in equations (2.120); this new determinant is then equal to the value of $\{E\}$ operating on y_1. In other words, we obtain

$$y_1 = \frac{\begin{vmatrix} te^{-t} & 4D & 3 \\ 4t^2 & 1 & -3D \\ 3\cos 2t & -5 & -2 \end{vmatrix}}{\begin{vmatrix} 2D^2 & 4D & 3 \\ D & 1 & -3D \\ 3D-2 & -5 & -2 \end{vmatrix}} \tag{2.121}$$

The two determinants appearing in equation (2.121) must now be evaluated. We may expand the numerator by *minors*, in which a series of simpler determinants multiply the elements of the row or column about which the expansion occurs. Expansion about the first column results in the following expression:

$$\begin{vmatrix} te^{-t} & 4D & 3 \\ 4t^2 & 1 & -3D \\ 3\cos 2t & -5 & -2 \end{vmatrix}$$

$$= \begin{vmatrix} te^{-t} & \cdot & \cdot \\ \cdot & 1 & -3D \\ \cdot & -5 & -2 \end{vmatrix}(-1)^{1+1} + \begin{vmatrix} \cdot & 4D & 3 \\ 4t^2 & \cdot & \cdot \\ \cdot & -5 & -2 \end{vmatrix}(-1)^{2+1}$$

$$+ \begin{vmatrix} \cdot & 4D & 3 \\ \cdot & 1 & -3D \\ 3\cos 2t & \cdot & \cdot \end{vmatrix}(-1)^{3+1} \tag{2.122}$$

$$= (1)\begin{vmatrix} 1 & -3D \\ -5 & -2 \end{vmatrix}te^{-t} + (-1)\begin{vmatrix} 4D & 3 \\ -5 & -2 \end{vmatrix}4t^2 + (1)\begin{vmatrix} 4D & 3 \\ 1 & -3D \end{vmatrix}3\cos 2t$$

Notice that each term in the expansion above is multiplied by a factor $(-1)^{i+j}$, where i is the number of the row and j is the number of the column of that element in the original determinant which corresponds to this term in the expansion. This factor of $(-1)^{i+j}$ guarantees that the expansion term will have the correct sign (+ or −) corresponding to its position in the original determinant. Cross-multiplication of the minor terms gives

$$\begin{vmatrix} te^{-t} & 4D & 3 \\ 4t^2 & 1 & -3D \\ 3\cos 2t & -5 & -2 \end{vmatrix} = e^{-t}(13t-15) - 60t^2 + 64t + 135\cos 2t \tag{2.123}$$

[Cross-multiplication of a determinant simply means that the elements of the determinant are multiplied along the diagonal directions; in an $n \times n$ determinant, the n products which result from diagonal multiplications in the (top, left) to (bottom, right) direction are added together, followed by the subtraction from this sum of the n products resulting from diagonal multiplications in the (bottom, left) to (top, right) direction. A 2×2 determinant has only two terms; for example,

$$\begin{vmatrix} A & B \\ C & D \end{vmatrix} = AD - BC \tag{2.124}$$

whereas a 3×3 determinant follows the general rule

$$\begin{vmatrix} A & B & C \\ D & E & F \\ G & H & I \end{vmatrix} = AEI + BFG + CDH - CEG - AFH - BDI \qquad (2.125)$$

that is, the number of terms is equal to twice the value of n.] The denominator of the ratio shown in equation (2.121) is then similarly evaluated; after simplification, we obtain

$$\begin{vmatrix} 2D^2 & 4D & 3 \\ D & 1 & -3D \\ 3D-2 & -5 & -2 \end{vmatrix} = -66D^3 + 28D^2 - 24D + 6 \qquad (2.126)$$

The differential operators in the expression (2.126) act on the unknown variable y_1. The original equation (2.121), after evaluation of the determinants has been accomplished, then becomes [using equations (2.123) and (2.126)]

$$-66\frac{d^3 y_1}{dt^3} + 28\frac{d^2 y_1}{dt^2} - 24\frac{dy_1}{dt} + 6y_1 = (13t - 15)e^{-t} - 60t^2 + 64t + 135 \cos 2t \qquad (2.127)$$

Equation (2.127) is the single equation with one unknown (y_1) which was sought; similar equations for y_2 and y_3 can be obtained using the same approach.

The method of determinants allows us to target only some of the response functions associated with a system for analysis; this luxury of selectivity is a primary benefit of this approach.

2.7 REVIEW

In summary, we have reviewed the following topics, facts, relationships, or concepts in this chapter.

- Differential equations allow us to represent the behavior of dynamic systems.
- We have focused on linear differential equations with constant coefficients; the physical systems thereby represented are assumed to be of the lumped-parameter, linear type, where the coefficients represent system parameters (e.g., mass, heat capacity, etc.)
- The solution y (the response of the system) to the linear differential equation

$$a_n\frac{d^n y}{dx^n} + a_{n-1}\frac{d^{n-1} y}{dx^{n-1}} + \cdots + a_1\frac{dy}{dx} + a_0 y = f(x) \qquad (2.1)$$

can be expressed as the sum of a transient solution and a steady-state solution:

$$y = y_{ts} + y_{ss} \qquad (2.2)$$

- The transient portion y_{ts} of the response function y is independent of the applied forcing function $f(x)$.
- Using operator notation, where the differential operator D is defined by

$$D \equiv \frac{d}{dx} \qquad (2.4)$$

one may write the characteristic equation for the system as

$$a_n D^n + a_{n-1} D^{n-1} + \cdots + a_1 D + a_0 = 0 \qquad (2.7)$$

the roots of which determine y_{ts}.

- One may identify the following types of roots of the characteristic equation with a corresponding contribution to y_{ts} as noted below:

Rule 1: real, unrepeated root r_1:

$$y_{ts_A} = C_1 e^{r_1 x} \qquad (2.8)$$

Rule 2: real root value r_2, repeated m times:

$$y_{ts_B} = (C_2 + C_3 x + C_4 x^2 + \cdots + C_{m+1} x^{m-1}) e^{r_2 x} \qquad (2.9)$$

Rule 3: complex roots r_3, $r_4 = a \pm jb$, not repeated:

$$y_{ts_C} = C_{m+2} e^{ax} \sin(bx + \phi) \qquad (2.10)$$

Rule 4: complex roots r_5, $r_6 = g \pm jh$, repeated p times:

$$
\begin{aligned}
y_{ts_D} = {}& C_{m+3}e^{gx}\sin\left(hx + \theta_1\right) \\
& + C_{m+4}xe^{gx}\sin\left(hx + \theta_2\right) + \cdots \\
& + C_{m+p}x^{p-1}e^{gx}\sin\left(hx + \theta_{p-1}\right)
\end{aligned}
\tag{2.14}
$$

so that

$$
y_{ts} = y_{ts_A} + y_{ts_B} + y_{ts_C} + y_{ts_D}
\tag{2.16}
$$

- As a result of the roots of the characteristic equation, one may evaluate the system response in terms of:
  (a) *Oscillatory behavior:* If any roots have nonzero imaginary portions, y_{ts} will include an oscillatory contribution.
  (b) *Stability:* If any root has a positive real portion, y_{ts} will include a term that will exponentially increase with time t (i.e., the system will be unstable).
- The steady-state solution y_{ss} depends on the applied forcing function $f(x)$; if the functional portion of $f(x)$ is such that—after a certain order derivative has been obtained—all higher-order derivatives either vanish or demonstrate the same functional dependence on x as lower-order derivatives, one may express y_{ss} as a linear combination of the functional portion of $f(x)$ and each of its functionally different derivatives, where each term is multiplied by an undetermined coefficient which is evaluated via substitution into the original differential equation involving y:

$$
y_{ss} = Af_v(x) + Bf_v'(x) + Cf_v''(x) + \cdots
\tag{2.31}
$$

- The application of multiple forcing functions $f_1(x), f_2(x), \ldots$, results in a total response function, the steady-state portion of which is equal to the sum of the steady-state solutions $y_{ss_1}, y_{ss_2}, \ldots$, which would result from the application of each individual forcing function:

$$
y_{ss} = y_{ss_1} + y_{ss_2} + \cdots + y_{ss_n}
\tag{2.50}
$$

for the case in which n forcing functions are applied to the system simultaneously. In other words, the system responds to each of the applied forcing functions and its total steady-state response is simply the sum of these individual responses.

- One may use Taylor's series, retaining only its first two terms, to obtain a linear approximation for a nonlinear function:

$$y(x) \simeq y(x_0) + \frac{dy}{dx}\bigg|_{x=x_0} (x - x_0) \qquad (2.55)$$

which is valid about the operating point x_0 for a narrow range Δx.

- Systems with multiple outputs can be described by simultaneous differential equations which can be evaluated by the method of common coefficients or by application of Cramer's rule.

EXERCISES

2.1. Write the following differential equations in operator notation, where

$$D \equiv \frac{d}{dx}$$

(a) $5\dfrac{d^4y}{dx^4} + \dfrac{d^3y}{dx^3} + 2\dfrac{dy}{dx} + 10 = 10e^{-t}$

(b) $4\dfrac{d^2y}{dx^2} + 4\dfrac{dy}{dx} - 5y + 18 = 16\sin 6t$

(c) $10\dfrac{d^5y}{dx^5} + 5\dfrac{d^4y}{dx^4} + 6\dfrac{d^2y}{dx^2} + 2\dfrac{dy}{dx} + y = 10x + 5e^{4x}$

2.2. Obtain the transient portion y_{ts} of the solution for the equation

$$10\frac{d^2y}{dx^2} + 6\frac{dy}{dx} + 4y = 10\sin 5x$$

Do not evaluate any unknown constants in y_{ts}.

2.3. Write the following equations in standard differential notation, where

$$D \equiv \frac{d}{dx}$$

(a) $(D^4 - 6D^2 + 2D)y = 6x$

(b) $(34D^6 + 21D^4 + 3D^3 + 2D^2 + D + 20)y = 6\sin 4x + 2e^{-3x}$

(c) $(3D^3 + 2D)y = 5x + 3\sin 3x$

2.4. The characteristic equation for a system has the following roots:

$$r_1 = -3, \quad r_3 = -2 + j7, \quad r_5 = -1 + j2, \quad r_7 = -4,$$
$$r_2 = 4, \quad r_4 = -2 - j7, \quad r_6 = -2 - j5, \quad r_8 = 8$$

(a) The eight roots above do not form the entire set of all 10 roots; identify the additional root values for r_9 and r_{10}.

(b) Write the full expression for the transient solution y_{ts} which corresponds to the full set of 10 roots.

(c) Is the system behavior stable or unstable? Explain your reasoning.

2.5. Given the expression

$$2\frac{d^2y}{dx^2} + 2\frac{dy}{dx} + y + 5 = 10e^{-x}$$

with the given initial conditions

$$y = 0$$
$$\frac{dy}{dx} = 0$$

at $x = 0$:

(a) Determine $y = y_{ts} + y_{ss}$.

(b) Draw a graphical description of y as a function of x.

2.6. Given the differential equation (in operator notation)

$$(3D + 5)y = 4\sin 4x \qquad \text{where } y = 0 \text{ at } x = 0$$

(a) Do you expect the transient portion y_{ts} of the solution y to oscillate? Explain your reasoning. (Do not evaluate the equation.)

(b) Describe the functional behavior of the steady-state portion y_{ss} of the solution that is to be expected. Explain your reasoning. (Do not evaluate the equation.)

(c) Evaluate the equation to obtain y_{ts} and y_{ss}. Do your results agree with your earlier expectations?

2.7. Evaluate the equation (in operator notation)

$$(2D^2 - 4D + 7)y = \sin 4x$$

to obtain the response $y(x)$; initial conditions are $y = 0$ and $dy/dx = 0$ at $x = 0$.

2.8. Evaluate the differential equation

$$3\frac{d^3y}{dx^3} + 2\frac{d^2y}{dx^2} + 2\frac{dy}{dx} + 4y = 18$$

to obtain the response $y(x)$; initial conditions are

$$y = 0$$

$$\frac{dy}{dx} = 2$$

$$\frac{d^2y}{dx^2} = 0$$

at $x = 0$.

2.9. Evaluate the expression

$$4\frac{dy}{dx} - 10y = 10x$$

to obtain $y(x)$; the initial condition is $y(0) = 0$.

2.10. A forcing function $f(x) = 4x^2$ is applied to a system; write the functional form of the steady-state solution y_{ss} for this case.

2.11. The mobility μ_I of ions in a semiconductor can be expressed as a function of temperature T by the formula (Rose et al., 1966)

$$\mu_I = A(T)^{3/2}$$

where A is a constant.

Obtain a linear approximation for $\mu_I(T)$ about the operating value $T_0 = 100°K$. Identify the slope and the intercept of the linear approximation.

2.12. Obtain the linear approximation (through the application of Taylor's series) for the function $f(x)$ about the operating point $x_0 = 2$, where $f(x)$ is given by the expression

$$f(x) = 4e^{3x}$$

Identify the slope and the intercept of the linear approximation.

2.13. The function $f(\theta)$ is given by the expression

$$f(\theta) = 5 \sin 2\theta + 4$$

(a) Determine the linear approximation (through the use of Taylor's series) for $f(\theta)$ about the operating point $\theta_0 = 0$.

(b) Determine the linear approximation for $f(\theta)$ about the operating point $\theta_0 = 90°$.

(c) Diagram the function $f(\theta)$, together with each of the linear approximations found in parts (a) and (b), versus the independent variable θ.

(d) Determine the effect of replacing the constant term (equal to 4) in the expression for $f(\theta)$ given above with a value equal to zero (0). Diagram this effect on the chart developed in part (c).

2.14. Consider the set of simultaneous equations

$$(3D^2 + 1)y_1 + (2D)y_2 + 4y_3 = 3 \sin 3x$$

$$30y_1 + (3D)y_2 + (3D)y_3 = 4x$$

$$(2D^2 + 2)y_1 + 2y_2 + 4y_3 = 3e^{-5x} + (D^2) \sin 2x$$

(a) Obtain an expression for $y_1(x)$ in which only $y_1(x)$ appears as a dependent variable.

(b) Obtain an expression for $y_3(x)$ in which only $y_3(x)$ appears as a dependent variable.

2.15. Use Cramer's rule to obtain an expression for $y_2(x)$ from the simultaneous equations

$$(D^2 + D + 2)y_1 \quad + \quad 3y_2 = 3\sin 4x$$

$$12y_1 + (D + 2)y_2 = 4e^{-2x} + 4x$$

2.16. Use Cramer's rule to obtain an expression for $y_1(x)$ from the simultaneous equations

$$2y_1 \quad + \quad (D)y_2 = 3\cos 2x$$

$$(5D)y_1 + (D + 1)y_2 = 2e^{-2x}$$

Furthermore, determine $y_1(x)$ from this expression; you may use any initial conditions of your choice.

2.17. Determine the function $y(x)$ that represents the solution to the differential equation

$$(3D^2 + D + 2)y = 3e^{-2x} + 4\cos 5x$$

where

$$y = 0, \quad \frac{dy}{dx} = 0,$$

at $x = 0$.

2.18. Determine the function $y(x)$ that represents the solution to the differential equation

$$(2D^2 + D - 3)y = 3e^{-2x} + 3x$$

where

$$y = 0, \quad \frac{dy}{dx} = 0$$

at $x = 0$. If $y(x)$ represents the response function of a system, is the system behavior stable or unstable? Explain your reasoning.

2.19. Consider the mechanical system shown in the diagram below. Two masses, M_1 and M_2, are connected together with a spring (with spring constant k_1); furthermore, the masses are also connected to fixed walls with two other springs (k_2 and k_3) and a mechanical dashpot or damper (with damping coefficient a). The applied force $f(t)$ is equal to a constant A.

(a) Use Newton's second law to develop a set of simultaneous differential equations that describe the behavior of this system.

(b) Use Cramer's rule to obtain an expression for $x_1(t)$ in which only $x_1(t)$ appears as a dependent variable.

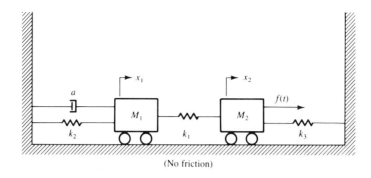

(No friction)

2.20. Given the two simultaneous equations (with zero initial conditions)

$$(5D)y_1 \ + \ 4y_2 = 2e^{-2x}$$

$$5y_1 + (3D)y_2 = 10x$$

where D is the differential operator d/dx.
(a) Obtain a single differential equation in which y_1 appears as the only dependent variable.
(b) Determine the transient portion y_{ss_1} of y_1.
(c) Determine the transient portion y_{ss_2} of y_2 and compare y_{ss_2} to y_{ss_1}.

2.21. Richardson (1960) proposed the following mathematical representation of the armament competition (or arms race) between two nations:

$$\frac{dx_1}{dt} = ax_2 - gx_1 + f_1$$

$$\frac{dx_2}{dt} = bx_1 - hx_2 + f_2$$

where

$x_1, x_2 \equiv$ armament levels (or the costs associated with such levels) for each nation

$a, b \equiv$ defense or reaction coefficients (reflecting the reaction of each nation to the size of the armament of its competitor)

$g, h \equiv$ fatigue or expense coefficients (denoting that economic constraints are imposed on the armament growth rate such that the rate decreases as the size of the armament increases)

$f_1, f_2 \equiv$ grievance coefficients (when positive) or goodwill coefficients (when negative)

(See also Saaty and Alexander, 1981.) The quantities f_1 and f_2 act as driving forces to either increase or decrease the rate of armament growth, depending on international conditions. Assume that $x_1 = x_2 = 10^5$ at time $t = 0$.

(a) Obtain a single second-order equation in which only x_1 appears as a dependent variable.

(b) Determine $x_1(t)$ for this system, given that

$$a = 2.0, \quad b = 1.0, \quad g = 1.0, \quad h = 5.0, \quad f_1 = 35.0, \quad f_2 = 40.0$$

(c) Determine $x_1(t)$ for this system, given that

$$a = 2.0, \quad b = 1.0, \quad g = 1.0, \quad h = 5.0, \quad f_1 = -5.0, \quad f_2 = -5.0$$

3

LAPLACE TRANSFORMATIONS

This is beyond anything which I could have imagined.

Dr. Watson, The Adventure of the Cardboard Box
by Sir Arthur Conan Doyle

3.1 OBJECTIVES

Upon completion of this chapter, the reader should be able to:

- Explain the concept of transformation as a mapping of a function from one plane of dependency into another plane.
- Define the general linear integral transformation of a function.
- Define the Laplace transformation of a function.
- State the sufficient conditions for the existence of the Laplace transform of a function.
- Obtain the Laplace transforms of several common functions (e.g., step function, ramp function, general polynomial in time t, exponential function in t, sinusoidal function).
- Obtain the Laplace transform of the nth-order derivative of a function in terms of the Laplace transform of the function itself and its lower-order ($<n$) derivatives evaluated at time t equal to zero.
- Explain the concept of an inverse Laplace transformation.
- Apply the Laplace transformation in the analysis of nth-order differential equations.
- Use the shift theorem expeditiously.
- Identify the advantages associated with the use of Laplace transformations in the solution of differential equations (e.g., the introduction

of initial conditions—immediately following the transformation operations—allows one to simplify the mathematical analysis during the early stages of the effort, in contrast to the classical approach, in which initial conditions are used near the end of the analysis).

* Apply the final-value and initial-value theorems to determine the extremum values of a function $x(t)$ at $t = 0$ and at $t = \infty$.

3.2 TRANSFORMATION OPERATORS

Now that we are familiar with the classical treatment of differential equations (Chapter 2), we will review the application of Laplace transformations to such equations. The Laplace transformation is part of the mathematical area known as *transformation calculus* (or *operational mathematics*). Through the use of this transformation, we will be able to develop the concept of the transfer function and its use in block diagrams of the system's organization. In addition, the Laplace transformation allows us to treat functions that increase without bound with increasing time t.

A transformation operator T, when applied to a function $F(t)$, results in a new form or "image" of this function. Examples of transformation operations include differentiation and integration:

$$D[F(t)] \equiv \frac{dF}{dt}$$

$$I[F(t)] \equiv \int_0^x F(t)\,dt$$

$$= f(x)$$

[Notice that the function $F(t)$ is transformed via the integration operation into a function of the "new" independent variable x; this type of "mapping" of a function from one plane of dependency (e.g., the t-plane) into another plane (e.g., the x-plane, in which one axis represents the independent variable x) will be of great value to us in developing conceptual insight into the mathematical description of a system. With such insight, we will be able to discern facts about the system and its behavior from the mathematical model without exhaustive analysis; as a result, we will be able to determine if an exhaustive analysis is needed or if the system itself is unsatisfactory and in need of modification.]

A transformation T is linear if it satisfies the principles of homogeneity and superposition; that is, a linear transformation must be of the form

$$T[C_1F_1(t) + C_2F_2(t)] = C_1T[F_1(t)] + C_2T[F_2(t)] \tag{3.1}$$

where C_1 and C_2 represent constant coefficients and where $F_1(t)$ and $F_2(t)$ are two functions of the independent variable t. We note that the principles of superposition and homogeneity are satisfied by such a transformation, as can be seen in the following cases for particular values of the constants C_1 and C_2:

$$T[C_1 F_1(t)] = C_1 T[F_1(t)] \tag{3.2}$$

(where C_2 has been set equal to zero) and

$$T[F_1(t) + F_2(t)] = T[F_1(t)] + T[F_2(t)] \tag{3.3}$$

(where both C_1 and C_2 have been set equal to unity).

Finally, we define the general linear integral transformation of a function $F(t)$ by the expression

$$T[F(t)] = \int_a^b K(t, s) F(t) \, dt$$
$$= f(s) \tag{3.4}$$

defined on the (finite or infinite) interval $a \leq t \leq b$ and where $K(t, s)$ is a function of the variable t and the parameter s. The result $f(s)$ of this transformation is the image of $F(t)$ in the s-plane or the s-domain; that is, we have transformed a function of the independent variable t into a form which is a function of the new independent variable s. To complete our description of this transformation, we need to specify both the class of functions to which $F(t)$ belongs and the range of the parameter s.

The Laplace transformation is a particular form of the general linear integral transformation T shown in equation (3.4), as will be demonstrated in the next section.

3.3 THE LAPLACE TRANSFORMATION

The function $K(t, s)$ is known as the *kernel* of the transformation given by equation (3.4). If we define the kernel as

$$K(t, s) \equiv e^{-st} \tag{3.5}$$

and if we further define the limits of the integration specified in equation (3.4) as extending over all values of t from zero to infinity, the general linear integral transformation T becomes the specific transformation known as the *Laplace transformation* $\mathscr{L}$:

$$
\begin{aligned}
\mathscr{L}[F(t)] &\equiv \int_0^\infty e^{-st} F(t)\, dt \\
&\equiv f(s) \\
&\equiv \text{Laplace transform of } F(t)
\end{aligned}
\tag{3.6}
$$

[Pierre Simon, Marquis de Laplace (1749–1827), and Augustin-Louis Cauchy (1789–1857) were early contributors to this area of operational calculus.]

Before we apply the Laplace transformation $\mathscr{L}$ to any particular functions or differential equations, we will state that the *sufficient* conditions for the existence of the Laplace transform of a function are that:

1. The function $F(t)$ is *sectionally continuous* in every finite interval in the range $t \geq 0$.
2. The function $F(t)$ is of *exponential order* as $t \to \infty$.

The first condition ensures that $F(t)$ is integrable for the interval $0 \leq t \leq U$ for some positive value of U and, together with the second condition, ensures that the Laplace integral will converge absolutely. [These conditions are sufficient but not necessary for the existence of the Laplace transform of a function; see, for example, Churchill (1958), upon which our treatment of the Laplace transform and its development in this chapter is largely based.]

[We note that a function $F(t)$ is sectionally continuous in a finite interval $a \leq t \leq b$ *if* the function is continuous within *each* of a finite number of subintervals and if $F(t)$ has finite limits as t approaches either of the end points of each subinterval from its interior. If $F(t)$ behaves in this manner, only "ordinary points of discontinuity" for this function will be found on the interval $a \leq t \leq b$ (i.e., discontinuities of finite size). Such sectionally continuous functions can be integrated over the entire interval $a \leq t \leq b$, where the integral equals the *sum* of the integrals of the continuous portions of $F(t)$ within the subintervals. This characteristic will be helpful to us in our consideration of Laplace transformations of nth-order derivatives.

If a constant λ exists such that the quantity

$$
e^{-\lambda t} |F(t)|
$$

is *bounded* for all values of $t > Q$, where Q is a finite number, then—as $t \to \infty$—$F(t)$ cannot grow faster than the product $M \cdot e^{\lambda t}$. Such a function

$F(t)$ is then said to be of exponential order as $t \to \infty$. Again, this property indicates that the Laplace transform of the function $F(t)$ will converge.]

Now that we have a working definition of the Laplace transform of a function [i.e., equation (3.6)], we may calculate the transforms of several functions which commonly appear in engineering analyses.

Example 3.1

Consider the function $F(t) = $ constant C. The Laplace transform of a constant is given by

$$
\begin{aligned}
\mathscr{L}[F(t)] &= \int_0^\infty e^{-st} F(t)\, dt \\
&= \int_0^\infty e^{-st} C\, dt \\
&= -\frac{C}{s} e^{-st}\Big|_0^\infty = 0 + \frac{C}{s} \\
&= \frac{C}{s}
\end{aligned}
\tag{3.7}
$$

Example 3.2

Consider the function $F(t) = Ct$, where C is a constant. The Laplace transform of this function is given by

$$
\begin{aligned}
\mathscr{L}[F(t)] &= \int_0^\infty e^{-st} F(t)\, dt \\
&= \int_0^\infty e^{-st} Ct\, dt
\end{aligned}
$$

which becomes, according to equation (3.2),

$$
\begin{aligned}
\mathscr{L}[F(t)] &= C \int_0^\infty e^{-st} t\, dt \\
&= -C\frac{t}{s} e^{-st}\Big|_0^\infty - \frac{C}{s}\int_0^\infty e^{-st}\, dt \\
&= -C\frac{t}{s} e^{-st}\Big|_0^\infty - C\left(-\frac{1}{s}\right)\left(-\frac{1}{s}\right) e^{-st}\Big|_0^\infty \\
&= C\left(-\frac{t}{s} - \frac{1}{s^2}\right) e^{-st}\Big|_0^\infty \\
&= 0 + \frac{C}{s^2} \\
&= \frac{C}{s^2}
\end{aligned}
\tag{3.8}
$$

(where we have used the technique of integration by parts to obtain the final value of the transformation).

Example 3.3

Consider the function $F(t) = Ce^{kt}$, where C and k are constants. The Laplace transform of this function is given by

$$\mathcal{L}[F(t)] = \int_0^\infty e^{-st}F(t)\,dt$$

$$= C\int_0^\infty e^{(k-s)t}\,dt \tag{3.9}$$

$$= \frac{C}{k-s}e^{(k-s)t}\Big|_0^\infty = -\frac{C}{k-s} = \frac{C}{s-k}$$

Example 3.4

Consider the step function $F(t)$, which behaves as follows:

$$F(t) = \begin{cases} 0 & \text{for } 0 \le t < k \\ C & \text{for } k \le t < \infty \end{cases} \tag{3.10}$$

where C is a constant, nonzero value. The Laplace transform of this function is obtained from an evaluation of the two subintervals defined by equation (3.10), in each of which $F(t)$ is continuous [i.e., we will use the fact that $F(t)$ is sectionally continuous]. As a result, we obtain

$$\mathcal{L}[F(t)] = \int_0^\infty e^{-st}F(t)\,dt$$

$$= \int_0^k e^{-st}(0)\,dt + C\int_k^\infty e^{-st}\,dt$$

$$= 0 - \frac{C}{s}e^{-st}\Big|_k^\infty \tag{3.11}$$

$$= \frac{C}{s}e^{-ks}$$

(where we have assumed that $s > 0$).

We can obtain the Laplace transformations of many other functions by using the definition of the transform, equation (3.6). Our analysis can become *more* efficient (and more useful) if we now introduce the Laplace transformations of derivatives.

3.4 THE LAPLACE TRANSFORM OF DERIVATIVES

We wish to be able to apply the Laplace transformation in our efforts to obtain quick solutions to differential equations. As a result, we need to be able to obtain the Laplace transform of the derivative(s) of a function.

Recall that the sufficient conditions for the existence of the Laplace transform of a function $F(t)$ are that (1) $F(t)$ is sectionally continuous, and (2) $F(t)$ is of exponential order as time t approaches infinity. In addition, recall that integration by parts (see Appendix B) allows us to express an integral in the form

$$\int u \, dv = uv - \int v \, du + C \qquad (3.12)$$

With these conditions for $F(t)$ and expression (3.12), we may calculate the Laplace transform of the first derivative $F'(t)$ of the function $F(t)$:

$$\mathscr{L}[F'(t)] = \int_0^\infty e^{-st} F'(t) \, dt$$
$$= e^{-st} F(t) \Big|_0^\infty + s \int_0^\infty F(t) e^{-st} \, dt \qquad (3.13)$$

The first term on the right side of equation (3.13) approaches zero as t approaches infinity if $F(t)$ is of exponential order [since s can be set larger than the constant λ for which the quantity $e^{-\lambda t}|F(t)|$ is bounded for all values of $t > Q$]. Therefore,

$$\mathscr{L}[F'(t)] = [0 - e^0 F(0)] + s \int_0^\infty F(t) \, e^{-st} \, dt \qquad (3.14)$$

or

$$\mathscr{L}[F'(t)] = s\mathscr{L}[F(t)] - F(0) \qquad (3.15)$$

Notice that equation (3.15) allows us to replace, via Laplace transformation, the operation of differentiation with an equivalent set of *algebraic* operations (i.e., multiplication of the Laplace transform of the undifferentiated function by the variable s, followed by subtraction of the value of the function with t set equal to zero). This property of replacement provides the basis for our use of the Laplace transform in the analysis of differential equations.

We may extend equation (3.15) to the case of the second derivative of a function $F(t)$ if the function $F(t)$ is sectionally continuous and if $F'(t)$ is continuous within the interval under investigation, and if both $F(t)$ and $F'(t)$ are of exponential order (these are not very restrictive conditions to place on the functions with which we will deal in physical systems). Then

$$\mathscr{L}[F''(t)] = s\mathscr{L}[F'(t)] - F'(0) \qquad (3.16)$$

or, with expression (3.15) for $\mathscr{L}[F'(t)]$,

$$\mathscr{L}[F''(t)] = s[s\mathscr{L}[F(t)] - F(0)] - F'(0)$$
$$= s^2\mathscr{L}[F(t)] - sF(0) - F'(0) \tag{3.17}$$

As Churchill (1958) notes, mathematical induction allows us to then extend this development to the nth-order derivative of a function $F(t)$, where the nth-order derivative $F^{(n)}(t)$ is sectionally continuous, the derivative of order $(n - 1)$ is continuous, and the function $F(t)$ and its first $(n - 1)$ derivatives $F'(t)$, $F''(t)$, ..., $F^{(n-2)}(t)$, $F^{(n-1)}(t)$ are of exponential order $e^{\lambda t}$ as t approaches infinity. The transform of $F^{(n)}(t)$ exists for $s > \lambda$ and is given by the expression

$$\boxed{\begin{aligned} \mathscr{L}[F^{(n)}(t)] &= s^n\mathscr{L}[F(t)] - s^{n-1}F(0) - s^{n-2}F'(0) \\ &\quad - \cdots - sF^{(n-2)}(0) - F^{(n-1)}(0) \end{aligned}} \tag{3.18}$$

In addition to providing us with the ability to analyze differential equations easily upon transformation, expression (3.18) can be used to obtain the Laplace transform of a function $F(t)$ more easily than with the use of the definition, equation (3.6), of the transform; this can be seen in the following examples.

Example 3.5

Consider the function $F(t) = 3 \sin 4t$.

$$F'(t) = 12 \cos 4t$$
$$F''(t) = -48 \sin 4t$$
$$= -16F(t)$$

Equation (3.18) then states that

$$\mathscr{L}[F''(t)] = s^2\mathscr{L}[F(t)] - sF(0) - F'(0)$$
$$= s^2\mathscr{L}[F(t)] - 0 - 12$$

However,

$$\mathscr{L}[F''(t)] = -16\mathscr{L}[F(t)]$$

or

$$\mathscr{L}[F(t)] = \frac{12}{s^2 + 16}$$

In general, for a function $F(t) = A \sin kt$, we have

$$\mathcal{L}[F''(t)] = As^2 \mathcal{L}[\sin kt] - s(A \sin 0) - kA \cos 0$$

$$= As^2 \mathcal{L}[\sin kt] - 0 - kA$$

but since

$$\mathcal{L}[F''(t)] = -k^2 \mathcal{L}[A \sin kt] = -k^2 \mathcal{L}[F(t)]$$

we then have

$$\mathcal{L}[A \sin kt] = \frac{Ak}{s^2 + k^2} \tag{3.19}$$

Example 3.6

Consider the function $F(t) = Ct^m$, where m is a known constant. We can determine the Laplace transform of $F(t)$ as follows:

$$F'(t) = mCt^{m-1}$$

$$F''(t) = m(m-1)Ct^{m-2}$$

$$\vdots \qquad \vdots$$

$$F^{(m)}(t) = (m!)Ct^0$$

$$= (m!)C$$

$$F^{(m+1)}(t) = 0$$

Application of equation (3.18) then gives the result [where the order n of the derivative is equal to $(m + 1)$]

$$\mathcal{L}[F^{(m+1)}(t)] = s^{m+1} \mathcal{L}[F(t)]$$

$$- s^m F(0) - s^{m-1} F'(0) - s^{m-2} F''(0)$$

$$- \cdots - F^{(m)}(0) \tag{3.20}$$

$$= s^{m+1} \mathcal{L}[F(t)] - 0 - 0 - \cdots - m!C$$

$$= s^{m+1} F(t) - m!C$$

However, since $F^{(m+1)}(t) = 0$, we also have

$$\mathcal{L}[F^{(m+1)}(t)] = 0$$

so that equation (3.20) becomes

$$\mathcal{L}[F(t)] = C\frac{m!}{s^{m+1}}$$

In general,

$$\mathcal{L}[F(t)] = \mathcal{L}[Ct^m]$$

$$= C\mathcal{L}[t^m]$$

$$= C\frac{m!}{s^{m+1}}$$

or

$$\mathcal{L}[t^m] = \frac{m!}{s^{m+1}} \tag{3.21}$$

3.5 INVERSE LAPLACE TRANSFORMATIONS

The Laplace transform maps a function $F(t)$, which can be described in the t-plane or the time domain, into an s-plane where it is described by its equivalent functional form $f(s)$, where

$$f(s) = \mathcal{L}[F(t)] \tag{3.22}$$

(see Figure 3.1).

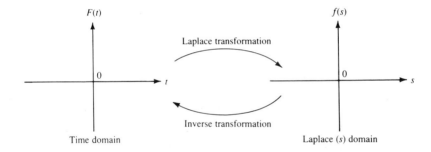

Figure 3.1   Mapping a function $F(t)$ into the Laplace domain.

Note that in Figure 3.1, an *inverse transformation* is indicated in which $f(s)$ is mapped *from* the s-plane *to* the t-domain, in which the function $F(t)$ appears. This inverse transformation operator $\mathcal{L}^{-1}$ is defined by the relationship

$$F(t) = \mathcal{L}^{-1}[f(s)] \tag{3.23}$$

together with equation (3.22).

The Lerch theorem (see Churchill, 1958) states that if two functions, $F_1(t)$ and $F_2(t)$, have the same Laplace transform $f(s)$, then

$$F_2(t) = F_1(t) + N(t)$$

where $N(t)$ is a *null function* defined by the expression

$$\int_0^T N(t) \, dt = 0 \tag{3.24}$$

for $T > 0$. As a result, the inverse transform $\mathcal{L}^{-1}$ is essentially unique since a null function is usually of no significance in our applications.

Explicit formulas for $\mathscr{L}^{-1}[f(s)]$ exist; one may also use tables of Laplace transforms to find $\mathscr{L}^{-1}[f(s)]$ in many cases (see Table 3.1).

After the Laplace transformation of a differential equation has been determined, the inverse transformation operator $\mathscr{L}^{-1}$ allows us to obtain the solution $F(t)$ in the t-domain to this differential equation *if* the solution $f(s)$ of the equivalent (Laplace-transformed) algebraic form of the equation has been obtained. Laplace transformation tables then provide the function $F(t)$ which corresponds to $f(s)$.

TABLE 3.1   Some Laplace Transforms of Value

Time Domain, $F(t)$	Laplace Domain, $f(s)$
Constant C	C/s
t^m	$\dfrac{m!}{s^{m+1}}$
Ce^{kt} ($C \equiv$ constant)	$\dfrac{C}{s-k}$
$C \sin kt$	$\dfrac{Ck}{s^2+k^2}$
$C \cos kt$	$\dfrac{Cs}{s^2+k^2}$
$\dfrac{d^nF}{dt^n}$	$s^nf(s) - s^{n-1}F(0) - s^{n-2}\dfrac{dF}{dt}\bigg|_{t=0}$ $\cdots - s\dfrac{d^{n-2}F}{dt^{n-2}}\bigg|_{t=0} - \dfrac{d^{n-1}F}{dt^{n-1}}\bigg|_{t=0}$
Unit impulse $\delta(t)$	1
$e^{at}F(t)$	$f(s-a)$
$F\left(\dfrac{t}{a}\right)$	$af(as)$
$1 - e^{-t/\tau}$	$\dfrac{1}{s(1+\tau s)}$
$\dfrac{s}{(s^2+k^2)^2}$	$\dfrac{1}{2k}t \sin kt$
$\dfrac{s^2}{(s^2+k^2)^2}$	$\dfrac{1}{2k}(\sin kt + kt \cos kt)$

Example 3.7

Consider the function $f(s) = m!/s^{m+1}$. From Example 3.6, we know that

$$F(t) = \mathscr{L}^{-1}\left[\frac{m!}{s^{m+1}}\right]$$

$$= t^m$$

Example 3.8

For the function $f(s) = k/(s^2 + k^2)$, we have

$$F(t) = \mathcal{L}^{-1}\left[\frac{k}{s^2 + k^2}\right]$$

$$= \sin kt$$

3.6 A SUBSTITUTION OF VALUE: THE SHIFT THEOREM

If a function $F(t)$—the Laplace transform $f(s)$ of which is known—is multiplied by the function e^{at} (where a is a known constant), then the Laplace transform of the product $e^{at}F(t)$ is equal to $f(s - a)$. One simply substitutes the quantity $(s - a)$ for the variable s in $f(s)$ to obtain $\mathcal{L}[e^{at}F(t)]$. In brief, for

$$f(s) = \mathcal{L}[F(t)]$$

we have

$$f(s - a) = \mathcal{L}[e^{at}F(t)] \tag{3.25}$$

Example 3.9

For $F(t) = t^2$,

$$f(s) = \mathcal{L}[t^2]$$

$$= \frac{2}{s^3}$$

Expression (3.25) states that

$$\mathcal{L}[t^2 e^{at}] = \frac{2}{(s - a)^3}$$

For $a = 4$,

$$\mathcal{L}[t^2 e^{4t}] = \frac{2}{(s - 4)^3}$$

Example 3.10

For $F(t) = 4 \sin 3t$,

$$\mathcal{L}[4 \sin 3t] = \frac{12}{s^2 + 9}$$

For the function $4e^{2t} \sin (3t)$, equation (3.25) gives

$$\mathcal{L}[4e^{2t} \sin 3t] = \frac{12}{(s - 2)^2 + 9}$$

$$= \frac{12}{s^2 - 4s + 13}$$

Expression (3.25) significantly extends our ability to obtain the Laplace transformation of numerous functions in which e^{at} appears as a multiplicand.

We may now apply our knowledge of Laplace transformations in the analysis of differential equations.

3.7 APPLICATION TO DIFFERENTIAL EQUATIONS

Coughanowr and Koppel (1965) have suggested that one use a step-by-step approach which is similar to that described below in order to solve a differential equation with Laplace transformations.

1. Obtain the Laplace transformation of *each side* of the differential equation and equate these transforms; this is permissible since the equality of two functions implies the equality of their transforms; that is,

$$F_1(t) = F_2(t)$$

   means that

$$\mathscr{L}[F_1(t)] = \mathscr{L}[F_2(t)]$$

   (see the Lerch theorem).
2. Introduce the initial conditions for the equation and solve algebraically for the Laplace transform $f(s)$ of the unknown function $F(t)$.
3. Use the inverse transformation operator $\mathscr{L}^{-1}$ to obtain the function $F(t)$ in the time domain. Expansion by means of *partial fractions* (see Appendix C) may be necessary during this step.

Example 3.11

Consider the first-order differential equation

$$\frac{dx}{dt} + 2x = 5t \tag{3.26}$$

where a single forcing function $(5t)$ is applied to the system under analysis and where $x(t) = 0$ for $t = 0$.

1. Obtain the Laplace transformations of each side of the equation:

$$\mathscr{L}\left[\frac{dx}{dt}\right] = s\mathscr{L}[x] - x(0) \tag{3.27}$$

$$\mathscr{L}[2x] = 2\mathscr{L}[x] \tag{3.28}$$

$$\mathscr{L}[5t] = 5\mathscr{L}[t]$$

$$= \frac{5}{s^2} \tag{3.29}$$

Combining the results of equations (3.27) through (3.29), we have

$$\mathcal{L}\left[\frac{dx}{dt} + 2x\right] = \mathcal{L}[5t]$$

or, equivalently,

$$s\mathcal{L}[x] - x(0) + 2\mathcal{L}[x] = \frac{5}{s^2}$$

so that

$$(s + 2)\mathcal{L}[x] - x(0) = \frac{5}{s^2} \tag{3.30}$$

2. Using the given initial conditions

$$x(0) = 0$$

simplifies equation (3.30) to the form

$$(s + 2)\mathcal{L}[x] = \frac{5}{s^2}$$

or, solving for the Laplace transform,

$$\mathcal{L}[x] = \frac{5}{s^2(s + 2)} \tag{3.31}$$

3. Using partial-fraction expansion, we obtain

$$\frac{5}{s^2(s + 2)} = \frac{A}{s + 2} + \frac{B}{s} + \frac{C}{s^2} \tag{3.32}$$

We then clear the equation of its denominators in order to determine the unknown coefficients A, B, and C:

$$5 = As^2 + B(s + 2)s + C(s + 2)$$
$$= (A + B)s^2 + (2B + C)s + 2C \tag{3.33}$$

Equating the coefficients of the terms in s on each side of expression (3.33) produces three equations for our three unknown coefficients:

$$A + B = 0 \tag{3.34a}$$

$$2B + C = 0 \tag{3.34b}$$

$$2C = 5 \tag{3.34c}$$

or

$$C = 2.5$$

$$B = -1.25$$

$$A = 1.25$$

As a result, we have

$$\mathcal{L}[x(t)] = \frac{1.25}{s+2} - \frac{1.25}{s} + \frac{2.5}{s^2} \tag{3.35}$$

Inverse transformations can then be performed as follows:

$$1.25\mathcal{L}^{-1}\left[\frac{1}{s+2}\right] = 1.25e^{-2t} \tag{3.36}$$

$$-1.25\mathcal{L}^{-1}\left[\frac{1}{s}\right] = -1.25 \tag{3.37}$$

$$2.5\mathcal{L}^{-1}\left[\frac{1}{s^2}\right] = 2.5t \tag{3.38}$$

thereby producing the final result:

$$x(t) = 1.25(e^{-2t} - 1) + 2.5t \tag{3.39}$$

[Notice that the final term in equation (3.39) is linear in t—due to the applied forcing function, which is also linear in t.]

Example 3.12

Consider the second-order differential equation

$$3\frac{dx^2}{dt^2} + 2\frac{dx}{dt} + 4x = 2e^{-5t} \tag{3.40}$$

with the initial conditions

$$x(0) = 0$$

$$x'(0) = 0$$

(In this example, the distinct forcing function $2e^{-5t}$ is applied to the system under analysis. We expect the output [described by the solution $x(t)$ of equation (3.40)] or response of the system to include an exponentially damped term in t as a result of this forcing function.)

1. Obtain the Laplace transformations of each side of equation (3.40).

$$3\{s^2\mathcal{L}[x(t)] - sx(0) - x'(0)\} + 2\{s\mathcal{L}[x(t)] - x(0)\} + 4\mathcal{L}[x(t)] = \frac{2}{s+5} \tag{3.41}$$

2. Apply the given initial conditions and solve for $\mathcal{L}[x(t)]$.

$$(3s^2 + 2s + 4)\mathcal{L}[x(t)] = \frac{2}{s+5}$$

or

$$\mathcal{L}[x(t)] = \frac{2}{(s+5)(3s^2 + 2s + 4)} \tag{3.42}$$

3. Obtain $x(t)$ through inverse transformations. Identifying the factors of the denominator of the right side of equation (3.42), we have the roots of the polynomial forming this denominator given by

$$r_1, r_2 = \frac{-2 \pm [(2)^2 - (4)(3)(4)]^{1/2}}{2(3)}$$

$$= -\frac{1}{3} \pm \frac{(-44)^{1/2}}{6} \tag{3.43a}$$

$$= -\frac{1}{3} \pm j(1.106)$$

where $j \equiv (-1)^{1/2}$; that is, roots r_1 and r_2 form a pair of complex conjugates. In addition, there is the purely real root r_3 given by

$$r_3 = -5 \tag{3.43b}$$

Partial-fraction expansion then produces

$$\frac{2}{(3s^2 + 2s + 4)(s + 5)} = \frac{A}{s + 5} + \frac{B}{s - r_1} + \frac{C}{s - r_2} \tag{3.44}$$

[Notice that we could obtain the *functional form* of $x(t)$ without determining the particular values of the unknown coefficients A, B, and C, that is,

$$\mathcal{L}^{-1}\left[\frac{A}{(s + 5)}\right] = Ae^{-5t}$$

$$\mathcal{L}^{-1}\left[\frac{B}{(s - r_1)}\right] = Be^{-r_1 t}$$

and so on. Such a functional analysis may be sufficient in certain cases; for example, if we determine that a particular root has a positive, real portion, we would then know that the system response is unstable.] Clearing equation (3.44) of fractions, we obtain

$$2 = A(s - r_1)(s - r_2) + B(s + 5)(s - r_2) + C(s + 5)(s - r_1)$$

$$= A[s^2 - (r_1 + r_2)s + r_1 r_2] + B[s^2 + (5 - r_2)s - 5r_2] \tag{3.45}$$

$$+ C[s^2 + (5 - r_1)s - 5r_1]$$

$$= (A + B + C)s^2 + [-A(r_1 + r_2) + B(5 - r_2) + C(5 - r_1)]s$$

$$+ (Ar_1 r_2 - 5r_2 B - 5r_1 C)$$

By inserting the values for r_1 and r_2 into equation (3.45) and equating coefficients of like terms in S, we may determine the values of A, B, and C. One can see that this approach is rather tedious; another method of solution is as follows. Multiply equation (3.44) by the denominator $(s + 5)$ of the term with the unknown coefficient A and then set s equal to the root value (-5) of this factor:

$$\frac{2}{3s^2 + 2s + 4} = A + \frac{B(s + 5)}{s - r_1} + \frac{C(s + 5)}{s - r_2}$$

followed by $s = -5$:

$$\tfrac{2}{69} = A + 0 + 0$$

or

$$A \simeq 0.029 \tag{3.46}$$

To determine the value of B, we multiply equation (3.44) by $(s - r_1)$ and then set s equal to r_1:

$$\frac{1}{s - r_2} \frac{2}{s + 5} = \frac{A(s - r_1)}{s + 5} + B + \frac{C(s - r_1)}{s - r_2}$$

which becomes, for $s = r_1$,

$$\frac{1}{r_1 - r_2} \frac{2}{r_1 + 5} = B$$

or, with equation (3.43),

$$B = \frac{1}{-1.222 + j(5.159)} \tag{3.47}$$

We can clear the denominator of equation (3.47) of imaginary numbers as follows (see Appendix A):

$$\begin{aligned} B &= \frac{1}{-1.222 + j(5.159)} \left[\frac{-1.222 - j(5.159)}{-1.222 - j(5.159)} \right] \\ &= \frac{-1.222 - j(5.159)}{28.1086} \end{aligned} \tag{3.48}$$

The value of C is similarly determined by multiplying equation (3.44) by $(s - r_2)$ and then setting s equal to r_2; the result is

$$C = \frac{-1.222 + j(5.159)}{28.1086} \tag{3.49}$$

Notice that C is the complex conjugate of B and that r_1 is the complex conjugate of r_2; we will comment further on this result.

Let us introduce the notation

$$B = a + jb \tag{3.50}$$

$$C = a - jb \tag{3.51}$$

$$r_1 = X + jY \tag{3.52}$$

$$r_2 = X - jY \tag{3.53}$$

where

$$a = -\frac{1.222}{28.1086} \simeq -0.0435 \tag{3.54}$$

$$b = -\frac{5.159}{28.1086} \simeq -0.1835 \tag{3.55}$$

$$X = -0.333 \tag{3.56}$$

$$Y = 1.106 \tag{3.57}$$

Operating on each term of equation (3.44) with the inverse transformation then gives

$$\mathscr{L}^{-1}\left[\frac{A}{s+5}\right] = Ae^{-5t} \tag{3.58}$$

$$\mathscr{L}^{-1}\left[\frac{B}{s-r_1}\right] = Be^{r_1 t}$$
$$= Be^{-0.333+j1.106} \tag{3.59}$$

$$\mathscr{L}^{-1}\left[\frac{C}{s-r_2}\right] = Ce^{r_2 t}$$
$$= Ce^{-0.333-j1.106} \tag{3.60}$$

We can demonstrate an interesting result by using the notation of equations (3.50) through (3.57), as follows:

$$Be^{t(X+jY)} = (a+jb)e^{t(X+jY)}$$
$$Ce^{t(X-jY)} = (a-jb)e^{t(X-jY)}$$

However, we know that

$$e^{t(X+jY)} = e^{Xt}(\cos Yt + j \sin Yt) \tag{3.61}$$

so that

$$Be^{t(X+jY)} + Ce^{t(X-jY)} = (a+jb)e^{Xt}(\cos Yt + j \sin Yt)$$
$$+ (a-jb)e^{Xt}(\cos Yt - j \sin Yt) \tag{3.62}$$
$$= 2e^{Xt}(a \cos Yt - b \sin Yt)$$

Thus expressions (3.58) through (3.60) and (3.62) give

$$x(t) = Ae^{-5t} + 2e^{Xt}(a \cos Yt - b \sin Yt) \tag{3.63}$$

Note that the first term in equation (3.63) can be attributed directly to the applied forcing function identified in the original differential equation (3.40), whereas the second (oscillation) term is due to the nonzero imaginary component (Y) of the roots r_1 and r_2. These results are consistent with the discussion of Chapter 2. Furthermore, note the simplicity of the second term of equation (3.63) relative to the initial form of this term given in equation (3.62); this reduction in complexity, as expressed in equation (3.62), is due to the fact that a pair of complex conjugate roots (r_1, r_2) will produce a pair of complex conjugate coefficients (B, C) in the solution $x(t)$ so that such a simplification can be obtained.

To demonstrate that a pair of complex conjugate roots of a denominator $G(s)$, where

$$\frac{1}{G(s)} = \frac{1}{(s + k_1 + jk_2)(s + k_1 - jk_2)}$$

$$= \frac{A}{s + k_1 + jk_2} + \frac{B}{s + k_1 - jk_2} \tag{3.64}$$

will result in coefficients (A and B) which form a pair of complex conjugates, we may argue as follows. $G(s)$ must be purely real if $G(s)$ is to represent the differential operations which are performed upon a system response $x(t)$ of a physical system (i.e., the coefficients of the differential operators appearing in the original differential equation represent physical parameters of the system, as discussed in Chapter 2, and must therefore be purely real values). If the coefficients A and B are assumed to be general complex variables, we then have

$$A = a_1 + jb_1 \tag{3.65a}$$

$$B = a_2 + jb_2 \tag{3.65b}$$

(where we wish to demonstrate that

$$a_1 = a_2 \tag{3.66a}$$

$$b_1 = -b_2 \tag{3.66b}$$

so that A is the complex conjugate of B, that is,

$$A = B^* \tag{3.67}$$

where $B^* \equiv a_2 - jb_2$). Clearing equation (3.64) of fractions, we obtain

$$1 = A(s + k_1 - jk_2) + B(s + k_1 + jk_2) \tag{3.68}$$

or, according to equations (3.65),

$$1 = (a_1 + jb_1)(s + k_1 - jk_2) + (a_2 + jb_2)(s + k_1 + jk_2)$$
$$= [(a_1 s + a_1 k_1 + b_1 k_2) + (a_2 s + a_2 k_1 - b_2 k_2)] \tag{3.69}$$
$$+ j[(b_1 s + b_1 k_1 - a_1 k_2) + (b_2 s + b_2 k_1 + a_2 k_2)]$$

For the equality to be valid in equation (3.69), the imaginary portion of the expression must vanish:

$$0 = (b_1 s + b_1 k_1 - a_1 k_2) + (b_2 s + b_2 k_1 + a_2 k_2)$$

or

$$b_1 s + b_1 k_1 - a_1 k_1 = -(b_2 s + b_2 k_1 + a_2 k_2) \tag{3.70}$$

Equating similar terms in s, we find that

$$b_1 = -b_2$$

which, upon substitution into equation (3.70) then gives

$$a_1 = a_2$$

We have obtained equations (3.66), as desired. This result then allows us to express a rational function $x(s)$ in the form

$$x(s) = \frac{F(s)}{G(s)}$$

$$= \frac{F(s)}{(s + k_1 + jk_2)(s + k_1 - jk_2)} \tag{3.71}$$

where, for physical systems, $G(s)$ will reflect the *design* of the system and $F(s)$ will represent the *input* function (i.e., the forcing function expressed in the s-domain). The function $x(s)$ will then denote the response or the *output* of the system (expressed in the s-domain) which can be expected when a transfer function $G(s)^{-1}$ operates on the input $F(s)$ to the system. We will discuss the use of transfer functions in control theory in succeeding chapters.

Expanding $x(s)$ in partial fractions results in the expression

$$x(s) = \frac{F(s)}{G(s)}$$

$$= F_1(s) + \frac{a_1 + jb_1}{s + k_1 + jk_2} + \frac{a_1 - jb_1}{s + k_1 - jk_2} \tag{3.72}$$

where $F_1(s)$ is a series of fractions (in its most general form) due to the function $F(s)$. An inverse Laplace transformation then gives

$$x(t) = \mathcal{L}^{-1}[x(s)]$$

$$= \mathcal{L}^{-1}[F_1(s)] + \mathcal{L}^{-1}\left[\frac{a_1 + jb_1}{s + k_1 + jk_2}\right] \tag{3.73}$$

$$+ \mathcal{L}^{-1}\left[\frac{a_1 - jb_1}{s + k_1 - jk_2}\right]$$

Notice that

$$\mathcal{L}^{-1}\left[\frac{a_1 + jb_1}{s + k_1 + jk_2}\right] = (a_1 + jb_1)\mathcal{L}^{-1}\left[\frac{1}{s + k_1 + jk_2}\right]$$

$$= (a_1 + jb_1)e^{-(k_1 + jk_2)t} \tag{3.74}$$

Furthermore,

$$e^{(x+jy)t} = e^{xt}(\cos yt + j \sin yt) \tag{3.75}$$

We then have

$$\mathscr{L}^{-1}\left[\frac{a_1 + jb_1}{s + k_1 + jk_2}\right] + \mathscr{L}^{-1}\left[\frac{a_1 - jb_1}{s + k_1 - jk_2}\right]$$

$$= 2e^{-k_1 t}(a_1 \cos k_2 t + b_1 \sin k_2 t) \qquad (3.76)$$

[Of course, this result is consistent with equation (3.63) in Example 3.12.]

Consider the significance of the result above in expression (3.76): Second-order differential equations will often result in a pair of complex conjugate roots $(-k_1 \pm jk_2)$ appearing in the analysis (as is also true for nth-order differential equations, where $n > 1$); equations (3.72), (3.73), and (3.76) allow us to determine the functional behavior of the system in the time domain. If the real portion $-k_1$ of the roots is negative, the contributions given by equation (3.76) vanish exponentially with time. If $-k_1$ is positive, these terms will increase without limit with increasing time t, thereby leading to instability. [If $F_1(t)$—which equals $\mathscr{L}^{-1}[F_1(s)]$ and which represents the system response to

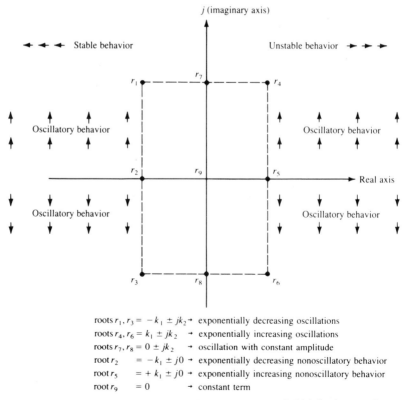

Figure 3.2 The physical significance of the root values of $G(s)$ in the complex s-plane.

the input $F(t)$—remains finite with increasing t, the system response is then completely stable for positive k_1 values.]

The imaginary term k_2 of the complex roots results in the *oscillatory* behavior of the system described in equation (3.76). Let us summarize the physical significance of the roots of $G(s)$ by considering the location of these roots in the complex s-domain (Figure 3.2).

3.8 FINAL-VALUE AND INITIAL-VALUE THEOREMS

We now introduce two theorems which will allow us to determine quickly the values of a function $X(t)$ at the boundary limits of $t = 0$ and $t = \infty$. The first relationship, known as the *initial-value theorem*, can be used to specify $X(t)$ at $t = 0$ if one has an expression for the Laplace transform [denoted by $x(s)$] of $X(t)$. This theorem states that

$$
\begin{aligned}
X(0) &= \lim_{t \to 0} X(t) \\
&= \lim_{s \to \infty} sx(s)
\end{aligned}
\tag{3.77}
$$

The second relationship is similar to expression (3.77) in that it focuses upon two limiting values—one of which is in the time domain and the other of which is in the s-domain:

$$
\begin{aligned}
X(\infty) &= \lim_{t \to \infty} X(t) \\
&= \lim_{s \to 0} sx(s)
\end{aligned}
\tag{3.78}
$$

Equation (3.78) is the mathematical form of the *final-value theorem*; it allows us to determine the value of $X(t)$ as t approaches infinity. This expression provides a mechanism to determine if a system's output $X(t)$ is stable *before* one performs an inverse Laplace transformation on $x(s)$ to obtain $X(t)$ itself. These theorems should be noted by the reader as *extremely useful* in systems analysis.

Example 3.13

Consider the Laplace transform given by

$$
x(s) = \frac{1}{2s^2 + 3s + 2}
$$

The final-value theorem gives $X(t = \infty)$:

$$X(\infty) = \lim_{t \to \infty} X(t)$$

$$= \lim_{s \to 0} sx(s)$$

$$= \lim_{s \to 0} \frac{s}{2s^2 + 3s + 2}$$

$$= 0$$

[The response $X(t)$ is bounded as t approaches infinity.]

Example 3.14

Consider the Laplace transform given by

$$x(s) = \frac{1}{s^4 + 3s^3 + 5s^2 + s}$$

The final-value theorem gives $X(t = \infty)$:

$$X(\infty) = \lim_{t \to \infty} X(t)$$

$$= \lim_{s \to 0} sx(s)$$

$$= 1$$

As in Example 3.13, $X(t)$ is bounded as t approaches infinity.

3.9 REVIEW

In summary, we have reviewed the following topics, facts, relationships, or concepts in this chapter.

- A transformation operator T, when applied to a function $F(t)$, results in a new form or image of this function; that is, $F(t)$ is mapped from one plane of dependency (the time or t-plane) into another plane of dependency.
- The Laplace transformation is a linear integral transformation defined by the expression

$$\mathscr{L}[F(t)] \equiv \int_0^\infty e^{-st} F(t) \, dt \qquad (3.6)$$

- Sufficient conditions for the existence of the Laplace transform of a function $F(t)$ are that:
  - (a) $F(t)$ is sectionally continuous in every finite interval in the range $t \geq 0$.
  - (b) $F(t)$ is of exponential order as $t \to \infty$.
- The Laplace transformation of the nth derivative $F^n(t)$ of a function $F(t)$ can be expressed in the form

$$\mathcal{L}[F^n(t)] = s^n \mathcal{L}[F(t)] - s^{n-1} F(0) - s^{n-2} F'(0)$$
$$- \cdots - sF^{(n-2)}(0) - F^{(n-1)}(0)$$

$$(3.18)$$

- An inverse Laplace transformation operator $\mathcal{L}^{-1}$ maps $f(s)$ from the Laplace s-plane to the time t-plane.
- The shift theorem allows us to extend our knowledge of the Laplace transform of a function $F(t)$,

$$f(s) \equiv \mathcal{L}[F(t)] \tag{3.22}$$

  to obtain easily the Laplace transform of this function when it is multiplied by an exponential function e^{at}, according to

$$f(s - a) = \mathcal{L}[e^{at}F(t)]$$

$$(3.25)$$

- Laplace transformations allow us to solve differential equations with ease, since we may use the initial condition(s) during the early phase of the analysis in order to simplify the algebraic relations in s which are equivalent to the differential equations. The procedure to be used is as follows:
  1. Obtain the Laplace transformation of each side of the differential equation and equate these transforms.
  2. Introduce the initial conditions and solve algebraically for the Laplace transform of the unknown dependent function.
  3. Use the inverse transformation operator to obtain the function in the time domain.
- The initial-value theorem

$$x(t = 0) = \lim_{s \to \infty} sx(s)$$

$$(3.77)$$

and the final-value theorem

$$x(t = \infty) = \lim_{s \to 0} sx(s) \qquad (3.78)$$

allow us to determine quickly the extremum values of $x(t)$.

EXERCISES

3.1. Explain the concept of transformation as a mapping of a function from one plane of dependency into another plane.

3.2. Determine the Laplace transform $\mathscr{L}[y(t)]$ of the function

$$y(t) = 4t^5$$

3.3. Determine the Laplace transform of the function

$$y(t) = 5e^{-4t} \sin 4t$$

3.4. Determine the Laplace transform of the function

$$y(t) = e^{-6t}(3t^2 + 5 \sin 5t)$$

3.5. Consider the following two simultaneous differential equations (where D represents the differential operator):

$$(3D)y_1 + \qquad 4y_2 = 5e^{-2t}$$

$$5y_1 + (3D)y_2 = 6t$$

(**a**) Obtain a single differential equation in which only y_2 appears as a dependent variable.

(**b**) Using Laplace transformations, determine the function $y_2(t)$. Use the initial conditions

$$y_2 = 0$$

$$\frac{dy_2}{dt} = 0$$

at $t = 0$.

(**c**) Is $y_2(t)$ consistent with stable behavior for a system? Why or why not?

3.6. Determine the function $y(t)$ if its Laplace transform is given by

$$\mathscr{L}[y(t)] = \frac{10}{s + 5} + \frac{5}{(s + 6)^2}$$

3.7. Consider the Laplace transform $f(s)$ given by

$$f(s) = \mathscr{L}[F(t)]$$

$$= \frac{15}{(s-8)^5}$$

Determine the function $F(t)$.

3.8. Determine the function $F(t)$ if its Laplace transform is given by

$$f(s) = \mathscr{L}[F(t)]$$

$$= \frac{2}{s-1} + \frac{15}{(s+3)^2+3}$$

3.9. Consider the differential equation

$$\frac{d^2x}{dt^2} + 5\frac{dx}{dt} - 2x = 4e^{-5t}$$

where $x(0) = 0$ at $t = 0$ and where $dx/dt = 0$ at $t = 0$. Use Laplace transformations to determine $x(t)$.

3.10. Consider the differential equation

$$3\frac{dx}{dt} + 3x = 4 + 3\cos 5t$$

where $x(0) = 2$ at $t = 0$ and where $dx/dt = 0$ at $t = 0$. Use Laplace transformations to determine $x(t)$.

3.11. Express the following differential equation in its equivalent Laplace-transformed form:

$$\frac{d^5x}{dt^5} + \frac{d^4x}{dt^4} + 2\frac{d^3x}{dt^3} + \frac{d^2x}{dt^2} + \frac{dx}{dt} + 5x = 3\sin 2t + 3e^{-5t}$$

3.12 The Laplace transform $x(s)$ of a function $X(t)$ is given by

$$x(s) = \frac{20}{3s^5 + 3s^3 + 2s + 5}$$

Determine the value of $X(t)$ as $t \to \infty$.

3.13. The Laplace transform $x(s)$ of a function $X(t)$ is given by

$$x(s) = \frac{36}{s(s+3)^2}$$

Determine the value of $X(t)$ at $t = 0$.

3.14. The Laplace transform $x(s)$ of a function $X(t)$ is given by

$$x(s) = \frac{28}{s^3 + 2s^2 + s}$$

Determine the value of $X(t)$ as $t \to \infty$.

3.15. For each of the following characteristic equations, evaluate y_{ts} with unspecified coefficients and phase factors. State if the system is stable or unstable in its behavior.

(a) $(s + 1)(s - 2)(s + 3) = 0$

(b) $(s^2 + 2s + 2)(s + 2) = 0$

(c) $s(s + 1)(s - 2)(s + 1)^2 = 0$

(d) $(s + 10)(s^2 - 2)(s + 2)^2 = 0$

(e) $(s - 5)(s - 0.1)(s - 0.01) = 0$

3.16. The Laplace transform $f(s)$ of a function $F(t)$ is given by the expression

$$f(s) = \frac{32}{(s^2 + 2s + 16)(s - 2)^2}$$

Determine $F(t)$.

3.17. Consider the differential equation

$$\frac{d^3y}{dt^3} + 3\frac{d^2y}{dt^2} + 8\frac{dy}{dt} + 24 = 2\sin 7t + 45$$

(a) Use the standard analytical techniques presented in Chapter 2 to evaluate $y(t)$. You may choose to use plotting and factoring methods to determine the roots of the characteristic equation. Assume zero initial conditions.

(b) Use Laplace transformations to evaluate $y(t)$. Compare your results to those obtained in part (a).

3.18. The Laplace transform $y(s)$ of a function $Y(t)$ is given by the expression

$$y(s) = \frac{1}{s - 1} + \frac{3}{s + 2} + \frac{1}{(s + 0.5)^2}$$

Determine $Y(t)$.

3.19. Consider the characteristic equation

$$s^2 + 6s + 7 = 0$$

(a) The response of the system is represented by $y(t)$. Determine the transient portion of $y(t)$ under the assumption of zero initial conditions.

(b) If the Laplace transform $f(s)$ of the applied forcing function $F(t)$ is given by

$$f(s) = \frac{2s + 5}{s^2(s + 2)^2}$$

determine $y(t)$ under the assumption of zero initial conditions.

4

NUMERICAL APPROXIMATION METHODS

It is an old maxim of mine that when you have excluded the impossible, whatever remains, however improbable, must be the truth.

Sherlock Holmes, The Adventure of the Beryl Coronet
by Sir Arthur Conan Doyle

4.1 OBJECTIVES

Upon completion of this chapter, the reader should be able to:

- Explain the need for numerical approximation methods in systems analysis.
- Recognize the need for numerical methods and the introduction of a computer in appropriate situations.
- Perform numerical analysis with the aid of a computer.
- Apply the Newton–Raphson iterative method to obtain an approximate value for a root of a function $f(x)$.
- Use the method of successive approximations to obtain a root of a function $f(x)$.
- Apply the secant method and regula falsi (method of false positions) to obtain a root of a function $f(x)$.
- Obtain quadratic factors of a polynomial via repeated application of Lin's method.
- Describe the limitations (particularly with respect to convergence) associated with each numerical approximation method described in this chapter.
- Analyze nonlinear ordinary differential equations via application of Euler's method or the Runge-Kutta methods.

4.2 THE NEED FOR NUMERICAL METHODS

In analyses of differential equations using the classical approach of Chapter 2 or the Laplace transformations of Chapter 3, one is confronted with the need to determine the factors (and the corresponding roots) of an nth-order polynomial of the form

$$p_n(x) = \sum_{i=0}^{n} a_i x^i$$

$$= a_n x^n + a_{n-1} x^{n-1} + a_{n-2} x^{n-2} + \cdots + a_1 x + a_0$$

(4.1)

(where, in the classical approach, the variable x is the differential operator D and, in the Laplace formulation, x is the Laplace variable s).

The quadratic formula allows us to easily determine the roots of a second-order expression; that is, for

$$p_2(x) = a_2 x^2 + a_1 x + a_0 \tag{4.2}$$

we may use

$$r_1, r_2 = \frac{[-a_1 \pm (a_1^2 - 4a_0 a_2)^{1/2}]}{2a_2} \tag{4.3}$$

if the coefficients a_2, a_1, and a_0 are known.

A third-order expression in x does not significantly increase our difficulty in analysis, since we may *estimate* the value of one of the three roots by considering only the two highest-order terms or the two lowest-order terms in $p_3(x)$, that is, approximate

$$p_3(x) = a_3 x^3 + a_2 x^2 + a_1 x + a_0 \tag{4.4}$$

by

$$p_3(x) \simeq a_3 x^3 + a_2 x^2 \tag{4.5a}$$

or by

$$p_3(x) \simeq a_1 x + a_0 \tag{4.5b}$$

thereby producing a (very) rough estimate of one root r_1 given by

$$r \simeq -\frac{a_2}{a_3} \tag{4.6a}$$

or

$$r \simeq -\frac{a_0}{a_1} \tag{4.6b}$$

One then plots $p_3(x)$ in the neighborhood of this estimated value r to find the actual value of r_1 (a real root). Division of $p_3(x)$ by the factor $(x - r_1)$ then produces the remaining quadratic factor from which the (real or complex) roots r_2 and r_3 can be determined in accordance with equation (4.3).

For $n \geq 4$, our difficulties in analysis become significant; for these higher-order polynomials, we will use numerical (approximation) methods to determine their linear and quadratic factors. *Some* of these methods will be described in the remainder of this chapter; the reader should refer to any one of the numerous texts (e.g., Arden and Astill, 1970; Carnahan et al., 1969) in numerical techniques which are available for additional approximation methods.

4.3 NEWTON–RAPHSON APPROXIMATION

Recall Taylor's series, equation (2.54), which we will write in the form

$$f(x) = f(x_0) + (x - x_0)f'(x_0) + (x - x_0)^2 \frac{f''(x_0)}{2!} + \cdots \tag{4.7}$$

where $f'(x_0)$ represents the first derivative of $f(x)$ evaluated at $x = x_0$, $f''(x_0)$ denotes the second derivative evaluated at x_0, and so on. Taylor's series can be used (Arden and Astill, 1970) as the basis for Newton's method (or the Newton–Raphson method) for determining the roots of a function, as follows. Retain only the first two terms in the series expansion of equation (4.7) (i.e., the linear or first-order approximation), so that

$$f(x) \simeq f(x_0) + (x - x_0)f'(x_0) \tag{4.8}$$

If x_1 is a root of the function $f(x)$, then

$$f(x_1) = 0$$
$$\simeq f(x_0) + (x_1 - x_0)f'(x_0) \tag{4.9}$$

so that we may solve equation (4.9) for x_1:

$$x_1 = x_0 - \frac{f(x_0)}{f'(x_0)} \tag{4.10}$$

If x_0 is an initial trial value for the root, equation (4.10) allows us to calculate a value x_1 which is a first-order improvement on the initial guess x_0. Repeated or iterative application of expression (4.10) then produces the $(k + 1)$th estimate of the root according to the relation

$$x_{k+1} = x_k - \frac{f(x_k)}{f'(x_k)} \tag{4.11}$$

Inclusion of the second-order term in Taylor's series (which we neglected to include in our first-order approximation) produces an additional term in expression (4.11) of the form

$$R = \frac{f''(x_k)}{2f'(x_k)}(x_{k+1} - x_k)^2 \tag{4.12}$$

which must decrease for such succeeding iteration if convergence to the actual root value is to occur.

The definition of a derivative can be expressed as

$$f'(x_0) = \lim_{x \to x_0} \frac{f(x) - f(x_0)}{(x - x_0)} \tag{4.13}$$

which we will approximate numerically by the *finite difference* expression

$$f'(x_0) \simeq \frac{f(x) - f(x_0)}{(x - x_0)} \tag{4.14}$$

or, for the $(k + 1)$th iteration,

$$f'(x_k) = \frac{f(x_{k+1}) - f(x_k)}{x_{k+1} - x_k} \tag{4.15}$$

If x_{k+1} is assumed to be the value of the root, $f(x_{k+1})$ must vanish, so that

$$f'(x_k) = -\frac{f(x_k)}{x_{k+1} - x_k} \tag{4.16}$$

or, solving for x_{k+1}:

$$x_{k+1} = x_k - \frac{f(x_k)}{f'(x_k)} \tag{4.17}$$

which is our iterative expression (4.11). Finite differences will aid us in the development of additional numerical techniques in Section 4.5.

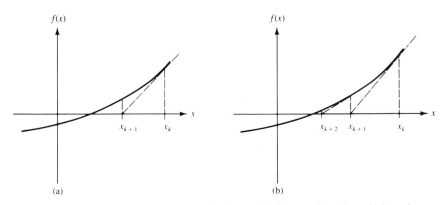

Figure 4.1 Newton-Raphson approximation method in graphical form (Adapted from Arden and Astill, 1970).

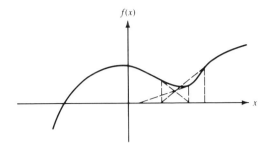

Figure 4.2   Failure to converge (Adapted from Arden and Astill, 1970).

A graphical description of this iterative technique (Figure 4.1a) shows that $f'(x_k)$ is the slope of the function $f(x)$ at $x = x_k$. The $(k + 2)$th iteration is shown in Figure 4.1b; as can be seen, each iteration results in a more accurate approximation to the actual root value of $f(x)$.

A word of caution: For some functions (e.g., see Figure 4.2), Newton's method may *not* result in convergence. Although this technique is valuable for many functions, we would be wise to extend our knowledge of iterative techniques for determining roots of functions—as we do in the remainder of this chapter.

4.4 METHOD OF SUCCESSIVE APPROXIMATIONS

A function $f(x)$ can be expressed as a combination of two functions and evaluated with respect to its root x as follows (Arden and Astill, 1970):

$$f(x) = g(x) - h(x)$$
$$= 0$$
$$(4.18)$$

where $g(x)$ can often be chosen to equal x itself, so that one obtains

$$x = h(x) \qquad (4.19)$$

which then provides an iteration formula

$$x_{k+1} = h(x_k) \qquad (4.20)$$

This technique, known as the *method of successive approximations*, is extremely useful *if* the iterative effort results in convergence to a root value. Convergence guarantees that a root has been determined, since in that case

$$x_{k+1} \cong x_k \qquad (4.21)$$

as $k \to \infty$, so that

$$x_{k+1} - h(x_k) = f(x_k)$$
$$= 0$$
$$(4.22)$$

It can be shown (Arden and Astill, 1970) that a sufficient condition for convergence is

$$|h'(x_k)| < 1$$

that is, the slope of $h(x)$ must be less than unity in its absolute magnitude.

4.5 SECANT METHOD AND REGULA FALSI

Finite differences, introduced in Section 4.3, allow us to express the first derivative of a function $f(x)$ with the approximation [based on the differential mean-value theorem; see Arden and Astill (1970) for a complete development]

$$f'(x_k) = \frac{f(x_{k+1}) - f(x_k)}{x_{k+1} - x_k} \tag{4.23}$$

which can be substituted into the first-order Taylor's series for $x_k = x_0$:

$$f(x) = f(x_0) + (x - x_0)\frac{f(x_1) - f(x_0)}{x_1 - x_0} \tag{4.24}$$

However, if x is a root of $f(x)$, then $f(x)$ vanishes, so that

$$0 = f(x_0) + (x - x_0)\frac{f(x_1) - f(x_0)}{x_1 - x_0}$$

or, solving for x,

$$x = x_0 - \frac{x_1 f(x_0) - x_0 f(x_0)}{f(x_1) - f(x_0)}$$

or, equivalently,

$$x = x_0\frac{f(x_1) - f(x_0)}{f(x_1) - f(x_0)} - \frac{x_1 f(x_0) - x_0 f(x_0)}{f(x_1) - f(x_0)}$$

$$= \frac{x_0 f(x_1) - x_1 f(x_0)}{f(x_1) - f(x_0)} \tag{4.25}$$

An iterative formula can then be generated from this result:

$$x_{k+1} = \frac{x_{k-1}f(x_k) - x_k f(x_{k-1})}{f(x_k) - f(x_{k-1})} \tag{4.26}$$

The iterative technique described by this expression is known as the *secant method* (or *Lin's method*) and may *not* converge to the desired root. However, if one requires that $f(x_k)$ and $f(x_{k-1})$ be opposite in sign (so that a root must lie between x_{k-1} and x_k, convergence is guaranteed; this special form of the secant method is known as the *method of false positions* or *regula falsi*. (Figure

4.3 describes this iterative technique graphically.) Another form of expression (4.26) is one in which the subscript notation indicates that x_{k-1} and x_k are on opposite sides of the root, one (denoted by x_L) to the left of the root and the other (x_R) to the right of the root, such that

$$x_{k+1} = \frac{x_R f(x_L) - x_L f(x_R)}{f(x_L) - f(x_R)} \tag{4.27}$$

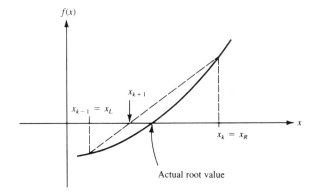

Figure 4.3 Graphical interpretation of the regula falsi method (Adapted from Arden and Astill, 1970).

After x_{k+1} is calculated, one determines $f(x_{k+1})$ and compares its sign to those of $f(x_L)$ and $f(x_R)$. If $f(x_{k+1})$ and $f(x_L)$ are identical in sign, the value of x_{k+1} replaces x_L in the next iteration; otherwise, x_{k+1} replaces x_R.

An example in which this method is applied will clarify its use.

Example 4.1

For turbulent flow of a fluid in a smooth pipe, the friction factor ν_f is related to the Reynolds number Re according to the expression

$$\left(\frac{1}{\nu_f}\right)^{1/2} = -0.4 + 1.74 \ln\left[\text{Re}\,(\nu_f)^{1/2}\right] \tag{4.28}$$

for turbulent flow of a fluid in a smooth pipe (Carnahan et al., 1969; Kay, 1963). Use the method of false positions to determine ν_f for Re equal to 10^4.

In this case, we seek the value of ν_f that will produce

$$f(\nu_f) = 0.4(\nu_f)^{1/2} + 1.74(\nu_f)^{1/2} \ln\left[\text{Re}\,(\nu_f)^{1/2}\right] + 1.0$$
$$= 0 \tag{4.29}$$

[i.e., we rewrite the original expression in a form $f(\nu_f)$ for which we may seek a root]. The regula falsi method is then used in accordance with the iterative expression

$$\nu_{f_k} = \frac{\nu_{f_R} f(\nu_{f_L}) - \nu_{f_L} f(\nu_{f_R})}{f(\nu_{f_L}) - f(\nu_{f_R})} \tag{4.30}$$

The use of a table is recommended for storage of iterative results. Arbitrarily chosen initial guesses for ν_{f_L} and ν_{f_R} are 10^{-8} and 1.0, respectively. [Calculate $f(\nu_{f_L})$ and $f(\nu_{f_R})$ in order to determine if they are opposite in sign.] Table 4.1 presents results for several iterations; note that the root is approached from the right [positive $f(\nu_k)$] after the first iteration is performed. This type of "one-direction" approach to the root is convenient for hand calculations; however, a simple computer program could be written to investigate a broader range of Reynolds numbers with greater accuracy.

TABLE 4.1   Regula Falsi Method Applied to Example 4.1, Equation (4.28): Iterative Results

Iteration k	ν_{L_k}	ν_{R_k}	$f(\nu_{L_k})$	$f(\nu_{R_k})$	ν_{f_k}	$f(\nu_{f_k})$
1	10^{-8}	1.0000000	1.4×10^{-4}	−14.626	9.6×10^{-6}	0.98275
2	9.6×10^{-6}	1.0000000	0.98275	−14.626	0.0672	−2.44180
3	9.6×10^{-6}	0.0672000	0.98275	−2.442	0.0193	−0.69328
4	9.6×10^{-6}	0.0192916	0.98275	−0.693	0.0113	−0.24747
5	9.6×10^{-6}	0.0113157	0.98275	−0.247	0.0090	−0.09652
6	9.6×10^{-6}	0.0090414	0.98275	−0.097	0.0082	−0.03901
7	9.6×10^{-6}	0.0082337	0.98275	−0.039	0.0079	−0.01599

4.6 LIN'S METHOD FOR QUADRATIC FACTORS

Returning to the problem of determining the roots of a polynomial

$$p_n(x) = a_n x^n + a_{n-1} x^{n-1} + \cdots + a_2 x^2 + a_1 x + a_0 \qquad (4.31)$$

we now introduce *Lin's method* (see Arden and Astill, 1970; Carnahan et al., 1969) for identifying quadratic factors of $p_n(x)$ (where, of course, $n > 2$). Once such a factor is determined, we may find the two corresponding roots of $p_n(x)$ via the quadratic formula. In addition, the remaining factor is a polynomial of order $(n - 2)$; repeated application of Lin's method allows us to identify the n roots of $p_n(x)$.

Assume that a quadratic factor of $p_n(x)$ is given by

$$x^2 + px + q$$

(where p and q must be determined). The polynomial can then be expressed as

$$p_n(x) = (x^2 + px + q)(b_n x^{n-2} + b_{n-1} x^{n-3} + \cdots + b_3 x + b_2) + b_1 x + b_0 \qquad (4.32)$$

If $(x^2 + px + q)$ is an exact quadratic factor of $p_n(x)$—as we assume—then the remainder term must vanish, that is,

$$b_1 = 0 \qquad (4.33a)$$

$$b_0 = 0 \qquad (4.33b)$$

Performing the multiplication in equation (4.32), we obtain

$$p_n(x) = b_n x^n + (b_{n-1} + b_n p)x^{n-1} + (b_{n-2} + b_{n-1}p + b_n q)x^{n-2}$$
$$+ \cdots + (b_2 + b_3 p + b_4 q)x^2 + (b_2 p + b_3 q)x + b_2 q \tag{4.34}$$

Equating coefficients of like terms in x in expressions (4.31) and (4.34) then gives $(n + 1)$ equations in $(n + 1)$ unknowns [where these unknowns are the $(n - 1)$ coefficients $b_n, b_{n-1}, \ldots, b_2$ together with p and q]:

$$a_n = b_n$$
$$a_{n-1} = b_{n-1} + b_n p$$
$$a_{n-2} = b_{n-2} + b_{n-1}p + b_n q$$
$$\vdots \qquad \vdots \tag{4.35}$$
$$a_2 = b_2 + b_3 p + b_4 q$$
$$a_1 = b_2 p + b_3 q$$
$$a_0 = b_2 q$$

Simultaneous evaluation of these $(n + 1)$ equations, beginning with initial trial values for p and q (i.e., p_0 and q_0), then produces estimates of the coefficients $b_n, b_{n-1}, \ldots, b_3, b_2$ together with new (more accurate) estimates for p and q.

A simple example will illustrate this technique.

Example 4.2

Determine the quadratic factors of the polynomial

$$p_4(x) = x^4 - 2x^3 + 1.25x^2 - 0.25x - 0.75 \tag{4.36}$$

(Carnahan et al., 1969).

The known coefficients are

$$a_4 = 1, \qquad a_3 = -2, \qquad a_2 = 1.25, \qquad a_1 = -0.25, \qquad a_0 = -0.75$$

We first identify the five $(= n + 1)$ equations which allow us to solve for these quadratic factors; from equations (4.35), we have

$$a_4 = b_4 \tag{4.37a}$$

$$a_3 = b_3 + b_4 p \tag{4.37b}$$

$$a_2 = b_2 + b_3 p + b_4 q \tag{4.37c}$$

$$a_1 = b_2 p + b_3 q \tag{4.37d}$$

$$a_0 = b_2 q \tag{4.37e}$$

With the given values for the coefficients a_i ($i = 1, 2, 3, 4$), these equations become the iterative formulas for b_4, b_3, and b_2, based on the kth iterative values for p and q:

$$b_{4_k} = 1 \tag{4.38}$$

$$b_{3_k} = -2 - b_{4_k} p_k \tag{4.39}$$

$$b_{2_k} = 1.25 - b_{3_k} p_k - q_k \tag{4.40}$$

with the remaining two expressions (4.37d) and (4.37e)

$$-0.25 = b_2 p + b_3 q$$

$$-0.75 = b_2 q$$

then allowing one to determine the $(k + 1)$ iterative values for p and q, that is,

$$q_{k+1} = \frac{-0.75}{b_{2_k}} \tag{4.41}$$

$$p_{k+1} = \frac{-0.25 - b_{3_k} q_k}{b_{2_k}} \tag{4.42}$$

[If one uses q_{k+1} in equation (4.42) instead of q_k, convergence to the final values for p and q will occur more rapdily.]

Table 4.2a presents a simple computer program which applies these iterative formulas to determine p and q; Table 4.2b presents the output of the program for initial trial values of $p = 0$ and $q = 0$. The unknown quantities p and q are determined to be equal to -1 and -0.75, respectively. Therefore,

$$p_4(x) = (x^2 - x - 0.75)(x^2 - x + 1) \tag{4.43}$$

where the second quadratic factor is determined by dividing $p_4(x)$ by the factor $(x^2 - x - 0.75)$. The roots of $p_4(x)$ are then given by

$$r_1 = 0.5 + j\left[\frac{(3)^{1/2}}{2}\right] \tag{4.44}$$

$$r_2 = 0.5 - j\left[\frac{(3)^{1/2}}{2}\right] \tag{4.45}$$

$$r_3 = -0.5 \tag{4.46}$$

$$r_4 = 1.5 \tag{4.47}$$

where $j \equiv (-1)^{1/2}$.

Example 4.3

(Based on Carnahan et al., 1969)   The equation of state for an ideal gas relates the pressure P, the volume V, and the temperature T according to

$$PV = nR_u T \tag{4.48}$$

where R_u is the universal gas constant (equal to 0.082054 liter $\cdot$ atm/mol $\cdot$ K) and n represents the number of moles of the gas under consideration. In such an approxima-

TABLE 4.2a   Sample Computer Program Using Lin's Method for Quadratic Factors

```
C       PROGRAM:  LIN'S METHOD FOR QUADRATIC FACTORS
C
C       ...A PROGRAM TO DETERMINE THE QUADRATIC ROOTS OF THE POLYNOMIAL:
C
C           F(X) = X**4 - 2*X**3 + 1.25*X**2 - 0.25*X - 0.75
C                = (X**2 + P*X + Q) * G(X)
C
C       THE 'STARTING VALUES' OF P AND Q, AND THE TOLERANCE AMIN ARE
C       CHOSEN BY THE USER.
C
        DATA A0,A1,A2,A3,A4/-0.75,-0.25,1.25,-2.00,1.00/
C
        OPEN(6,FILE='OUT.TXT')
C
        WRITE(*,*)    '  INITIAL VALUES OF P?  Q? '
        READ(*,*)     P, Q
        WRITE(*,*)    '  TOLERANCE LIMIT? '
        READ(*,*)     AMIN
        WRITE(6,100)  P, Q, AMIN
C
        B4 = A4
C
        DO 10 I = 1, 100
            B3 = A3 - A4*P
            B2 = A2 - (A3 - A4*P)*P - A4*Q
            P2 = (A1 - Q*B3)/B2
            Q2 = A0/B2
            WRITE(6,110) I, P2, Q2, B4, B3, B2
C
            IF ( ABS(P2-P) .LE. AMIN .AND. ABS(Q2-Q) .LE. AMIN) STOP
C
            P = P2
            Q = Q2
     10 CONTINUE
C
        CLOSE(6,STATUS='KEEP')
C
    100 FORMAT(//,T4,'LIN''S METHOD FOR QUADRATIC FACTORS: P = ',F5.2,
       1        ' Q = ',F5.2,' AMIN = ',F6.4,//,T4,'ITERATION',6X,'P',
       2_       9X,'Q',8X,'B4',6X,'B3',6X,'B2',/)
    110 FORMAT(6X,I3,6X,F7.4,3X,F7.4,3X,F5.3,3X,F6.3,3X,F5.3)
C
        END
```

tion, one assumes that the gas particles do not interact with one another except via occasional collisions.

A more accurate approximation to the behavior of real gases was proposed by van der Waals in 1873 (see Wark, 1977): The volume occupied by the particles is assumed to be equal to a constant b (known as the covolume of the particles); in addition, the intermolecular attractive forces are interpreted as an effective pressure equal to a/v^2, where a is a constant and v is the specific volume (i.e., the volume per unit mass or V/m, where m is the mass of the gas). If one introduces the specific gas constant R, defined by the relation

$$R \equiv \frac{R_u}{M} \tag{4.49}$$

TABLE 4.2b   Sample Output for Program Given in Table 4.2a[a]

LIN'S METHOD FOR QUADRATIC FACTORS: P =   .00 Q =   .00 AMIN =   .0001

ITERATION	P	Q	B4	B3	B2
1	-.2000	-.6000	1.000	-2.000	1.250
2	-.8926	-.5034	1.000	-1.800	1.490
3	-1.0556	-.9805	1.000	-1.107	.765
4	-.9533	-.6080	1.000	-.944	1.234
5	-1.0305	-.8719	1.000	-1.047	.860
6	-.9755	-.6679	1.000	-.970	1.123
7	-1.0172	-.8165	1.000	-1.024	.919
8	-.9866	-.7030	1.000	-.983	1.067
9	-1.0097	-.7868	1.000	-1.013	.953
10	-.9926	-.7233	1.000	-.990	1.037
11	-1.0055	-.7705	1.000	-1.007	.973
12	-.9958	-.7349	1.000	-.995	1.021
13	-1.0031	-.7615	1.000	-1.004	.985
14	-.9977	-.7415	1.000	-.997	1.012
15	-1.0017	-.7564	1.000	-1.002	.991
16	-.9987	-.7452	1.000	-.998	1.006
17	-1.0010	-.7536	1.000	-1.001	.995
18	-.9993	-.7473	1.000	-.999	1.004
19	-1.0006	-.7520	1.000	-1.001	.997
20	-.9996	-.7485	1.000	-.999	1.002
21	-1.0003	-.7511	1.000	-1.000	.998
22	-.9998	-.7491	1.000	-1.000	1.001
23	-1.0002	-.7506	1.000	-1.000	.999
24	-.9999	-.7495	1.000	-1.000	1.001
25	-1.0001	-.7504	1.000	-1.000	1.000
26	-.9999	-.7497	1.000	-1.000	1.000
27	-1.0001	-.7502	1.000	-1.000	1.000
28	-1.0000	-.7498	1.000	-1.000	1.000
29	-1.0000	-.7501	1.000	-1.000	1.000
30	-1.0000	-.7499	1.000	-1.000	1.000
31	-1.0000	-.7501	1.000	-1.000	1.000
32	-1.0000	-.7500	1.000	-1.000	1.000
33	-1.0000	-.7500	1.000	-1.000	1.000

[a] Initial trials values of $p = 0$ and $q = 0$, with AMIN equal to 0.0001.

where M is the molar mass of the gas, the ideal gas equation of state becomes

$$Pv = RT \tag{4.50}$$

The van der Waals equation of state is

$$\left(P + \frac{a}{v^2}\right)(v - b) = RT \tag{4.51}$$

Given a set of P, T, a, and b values, we may use Lin's method to determine the values of v that satisfy equation (4.51).

Rewriting equation (4.51) as

$$f(v) = Pv^3 + (-RT - Pb)v^2 + av - ab = 0 \tag{4.52}$$

expresses $f(v)$ in the form of a polynomial, that is,

$$f(v) = a_3 v^3 + a_2 v^2 + a_1 v + a_0$$
$$= 0$$

(4.53)

If we assume that the quadratic factor of this polynomial can be expressed as $(v^2 + pv + q)$, then

$$f(v) = (v^2 + pv + q)(b_3 v + b_2)$$
$$= 0$$

(4.54)

or

$$f(v) = b_3 v^3 + (pb_3 + b_2)v^2 + (b_2 p + b_3 q)v + b_2 q$$

(4.55)

[Note that the pressure P is not equal to the unknown coefficient p (i.e., $P \neq p$).]

Equating coefficients of like terms in v in equations (4.52), (4.53), and (4.55), we obtain the simultaneous iterative equations

$$b_{3_k} = a_3 = P$$

(4.56a)

$$b_{2_k} = a_2 - b_{3_k} p_k$$
$$= (-RT - Pb) - b_{3_k} p_k$$

(4.56b)

$$q_{k+1} = \frac{a_0}{b_{2_k}} = \frac{-ab}{b_{2_k}}$$

(4.56c)

$$p_{k+1} = \frac{a_1 - b_{3_k} q_k}{b_{2_k}}$$
$$= \frac{a - b_{3_k} q_k}{b_{2_k}}$$

(4.56d)

Table 4.3a presents a listing for a computer program which uses equations (4.56) to determine p and q for a range of pressures and temperatures as specified by the user; the user may also choose the incremental changes in pressure and temperature, together with the accuracy desired [in terms of $(p_{k+1} - p_k)$ and $(q_{k+1} - q_k)$].

The van der Waals constants a and b must also be specified by the user for the particular gas under investigation. Finally, the maximum number of iterations that are to be performed for a specific set of pressure–temperature values must also be specified by the user to ensure that the evaluation is reasonably bounded.

Table 4.3b presents the output of the program for benzene and ammonia which are investigated for the pressure range 1 to 5 atm (in 1-atm increments) and for the temperature range 100 to 500° K (in 100° K increments). [Of course, these (P, T) combinations are only examples of the wide range of (P, T) sets that can be considered with this program and Lin's method.]

TABLE 4.3a   **Sample Computer Program Using Lin's Method for Quadratic Factors to Evaluate the Roots of the van der Waal's Equation of State for an Imperfect Gas**

```
C     PROGRAM:  VAN DER WAALS EQUATION
C
C     ...A PROGRAM TO DETERMINE THE ROOTS OF THE VAN DER WAALS
C     EQUATION OF STATE FOR AN IMPERFECT GAS. THE USER SPECIFIES THE
C     RANGES OF PRESSURE (P) AND TEMPERATURE (T) TO BE INVESTIGATED,
C     AS WELL AS THE VAN DER WAALS CONSTANTS a AND b FOR THE PARTICULAR
C     GAS UNDER CONSIDERATION.
C
      CHARACTER*14 VAPOR
C
C     GAS CONSTANT R IS GIVEN IN LITER-ATM/MOLE-K
C
      DATA R / 0.082054 /
C
      OPEN(6,FILE='OUT.TXT')
C
C     INPUT INITIAL DATA
C
      WRITE(*,*)  '  PRESSURE RANGE:  PMIN? PMAX? DELP? '
      READ(*,*)   PMIN, PMAX, DELP
      WRITE(*,*)  '  TEMPERATURE RANGE:  TMIN? TMAX? DELT? '
      READ(*,*)   TMIN, TMAX, DELT
      WRITE(*,*)  '  VAN DER WAALS CONSTANTS:  a?  b? '
      READ(*,*)   A, B
      WRITE(*,*)  '  VAPOR? '
      READ(*,130) VAPOR
      WRITE(*,*)  '  ACCURACY OF THE ROOT APPROXIMATION: AMIN? '
      READ(*,*)   AMIN
      WRITE(*,*)  '  MAXIMUM # OF ITERATIONS PER (T,P) SET: IMAX? '
      READ(*,*)   IMAX
C
      WRITE(6,100) IMAX, AMIN, A, B, VAPOR
C
      A1   = A
      A0   = -A * B
      P    = PMIN
      T    = TMIN
      IP   = IFIX( (PMAX-PMIN)/DELP + 1. )
      IT   = IFIX( (TMAX-TMIN)/DELT + 1. )
      INC  = MIN0(IP,IT)
C
C     INITIAL VALUES FOR THE QUADRATIC COEFFICIENTS, PP AND Q, ARE
C     SET EQUAL TO ZERO
C
      PP = 0.
      Q  = 0.
C
      DO 10 J = 1, INC
         ITER = IMAX
         N    = 0
         B3   = P
         A2   = -P*B - R*T
C
         DO 20 I = 1, ITER
            B2 = A2 - B3*PP
            Q2 = A0/B2
            P2 = (A1 - B3*Q)/B2
C
            IF( ABS(P2-PP) .GT. AMIN .OR. ABS(Q2-Q) .GT. AMIN) THEN
               PP = P2
```

TABLE 4.3a   Sample Program (continued)

```
                Q   = Q2
                N   = N + 1
           ELSE
                ITER = I - 1
           END IF
C
   20   CONTINUE
C
           R3  = -B2/B3
           RAD = P2**2 - 4.*Q2
C
C    DETERMINE REAL OR COMPLEX ROOTS
C
           IF (RAD .LT. 0.0) THEN
                RADI = SQRT(-RAD) / 2.
                RADR = -P2 / 2.
                WRITE(6,110) T, P, N, R3, RADR, RADI
           ELSE
                R1 = ( -P2 + SQRT(RAD) ) / 2.
                R2 = ( -P2 - SQRT(RAD) ) / 2.
                WRITE(6,120) T, P, N, R3, R1, R2
           END IF
C
        P = P + DELP
        T = T + DELT
   10 CONTINUE
C
     CLOSE(6,STATUS='KEEP')
C
  100 FORMAT(//,T4,'VAN DER WAAL''S EQUATION...',//,T7,'CONSTRAINTS:',
     1       ' MAX # OF ITERATIONS PER (T,P) SET = ',I4,',   ACCURACY ',
     3       '= ',F6.4,//,T7,'VAN DER WAAL''S CONSTANTS: A = ',F7.4,
     3       ',   B = ',F6.4,'  (',A14,')',/)
  110 FORMAT(/,T4,'T = ',F6.2,',   P = ',F6.2,',   ITER = ',I4,/,T8,
     1       'REAL ROOT = ',F7.5,',   COMPLEX ROOTS = ',F7.5,' +,- j',
     2       F7.5)
  120 FORMAT(/,T4,'T = ',F6.2,',   P = ',F6.2,',   ITER = ',I4,/,T8,
     1       'REAL ROOTS:  R1 = ',F7.5,',   R2 = ',F7.5,',   R3 = ',F7.5)
  130 FORMAT(2X,A14)
C
        END
```

The constants a and b for various gases are as follows:

Gas	a (liter^2 · atm/mol^2)	b (liter/mol)
Acetylene (C_2H_2)	4.410	0.0510
Air	1.358	0.0364
Ammonia (NH_3)	4.233	0.0373
Benzene (C_6H_6)	18.63	0.1181
Carbon dioxide (CO_2)	3.643	0.0427
Ethylene (C_2H_4)	4.563	0.0574
Helium (He)	0.0341	0.0234
Water (H_2O)	5.507	0.0304

Source: Wark (1977).

TABLE 4.3b Sample Output of Program Given in Table 4.3a: Values for Benzene and Ammonia

---

```
VAN DER WAAL'S EQUATION...

   CONSTRAINTS: MAX # OF ITERATIONS PER (T,P) SET = 1000,  ACCURACY = .0001
   VAN DER WAAL'S CONSTANTS: A = 4.2330,  B = .0373  (AMMONIA      )


T = 100.00,  P =  1.00,  ITER =    4
   REAL ROOTS:  R1 = 7.69530,  R2 = .50694,  R3 = .04047

T = 200.00,  P =  2.00,  ITER =    3
   REAL ROOTS:  R1 = 7.97866,  R2 = .21881,  R3 = .04522

T = 300.00,  P =  3.00,  ITER =    2
   REAL ROOTS:  R1 = 8.06861,  R2 = .11947,  R3 = .05460

T = 400.00,  P =  4.00,  ITER =    2
   REAL ROOT = 8.11285,  COMPLEX ROOTS = .06492 +,- j .02551

T = 500.00,  P =  5.00,  ITER =    2
   REAL ROOT = 8.13916,  COMPLEX ROOTS = .05177 +,- j .03464

VAN DER WAAL'S EQUATION...

   CONSTRAINTS: MAX # OF ITERATIONS PER (T,P) SET = 1000,  ACCURACY = .0001
   VAN DER WAAL'S CONSTANTS: A = 18.6300,  B = .1181  (BENZENE      )


T = 100.00,  P =  1.00,  ITER = 1000
   REAL ROOTS:  R1 = 4.27373,  R2 = 4.11945,  R3 = .12497

T = 200.00,  P =  2.00,  ITER =    7
   REAL ROOTS:  R1 = 7.01863,  R2 = 1.17100,  R3 = .13385

T = 300.00,  P =  3.00,  ITER =    4
   REAL ROOTS:  R1 = 7.50952,  R2 = .66767,  R3 = .14627

T = 400.00,  P =  4.00,  ITER =    3
   REAL ROOTS:  R1 = 7.73015,  R2 = .42644,  R3 = .16686

T = 500.00,  P =  5.00,  ITER =    3
   REAL ROOT = 7.85636,  COMPLEX ROOTS = .23357 +,- j .03817
```

4.7 NONLINEAR ORDINARY DIFFERENTIAL EQUATIONS: EULER'S METHOD AND THE RUNGE–KUTTA METHODS*

Taylor's series, equation (2.54), enabled us to obtain a linear approximation of a nonlinear relationship. We now extend our ability to analyze nonlinear ordinary differential equations (which describe the behavior of physical systems) through the use of numerical approximations. [Our discussion is

---

* The material in this section can be omitted in those courses which are under difficult time constraints; we wish to emphasize, however, that the reader should become familiar with this material if time allows.

based on the treatment given by Carnahan et al. (1969), together with that of Arden and Astill (1970).]

Although we will restrict our attention to systems with linear behavior in the remainder of this work, we seek to demonstrate that nonlinear equations can be numerically evaluated by many well-known methods, such as Euler's method and the Runge–Kutta methods. Such numerical approximation methods can also be applied to the analysis of linear ordinary differential equations.

An nth-order ordinary differential equation has the form

$$F\left(x, y, \frac{dy}{dx}, \frac{d^2y}{dx^2}, \ldots, \frac{d^ny}{dx^n}\right) = 0 \qquad (4.57)$$

that is, it is a function of x, y, and the first n derivatives of y with respect to x.

We restrict our attention to those equations in which only total derivatives appear (no partial derivatives), that is, ordinary differential equations. [In *lumped-parameter* systems (to which we are restricting ourselves in this text), one assumes that the systems can be described as sets of finite elements. Partial differential equations often arise in the descriptions of *continuous* systems.] We seek a function $y(x)$ that satisfies equation (4.57). A unique $y(x)$ can be determined if n conditions are known; if all n conditions are specified at a particular value x_0 of the independent variable x, the problem is known as an initial-value problem. Otherwise, if the n conditions are distributed among different values of x (e.g., x_0, x_1, x_2), we are considering a boundary-value problem.

The introduction of $(n - 1)$ new variables allow us to write an nth-order differential equation of the type described by expression (4.57) as a system of n first-order equations, for example, the second-order equation

$$x^2\frac{d^2y}{dx^2} - xy\frac{dy}{dx} + 4y = 0 \qquad (4.58)$$

If we define

$$z \equiv \frac{dy}{dx} \qquad (4.59)$$

$$\rightarrow \qquad \frac{dz}{dx} = \frac{d^2y}{dx^2} \qquad (4.60)$$

we may then write

$$z - \frac{dy}{dx} = 0 \qquad (4.61a)$$

$$x^2\frac{dz}{dx} - xyz + 4y = 0 \qquad (4.61b)$$

The two first-order equations (4.61) are equivalent to the second-order equation (4.58). One could then use an iterative procedure in which initial trial values for x and y (at $x = x_0$) produce a value for z at x_0 from equation (4.61b); substitution of this value for z in equation (4.61a) then produces an estimate of the slope dy/dx of y at x_0. A linear approximation for y, for example,

$$y(x_0 + h) = y(x_0) + \frac{dy}{dx}\bigg|_{x_0} h \qquad (4.62)$$

then gives an estimate of y at $x = x_0 + h$. Iterations continue until the entire interval $(x_0 \leq x \leq x_{\text{final}})$ of interest has been investigated.

Most higher-order systems can be similarly analyzed as a set of first-order equations (Carnahan et al., 1969); therefore, we will focus on the treatment of such *first-order* equations of the type described by

$$F\left(x, y, \frac{dy}{dx}\right) = 0 \qquad (4.63)$$

or

$$f(x, y) = \frac{dy}{dx} \qquad (4.64)$$

It is usually not possible to evaluate *analytically* an expression such as (4.64) in order to determine $y(x)$; however, numerical approximation techniques do allow us to investigate $y(x)$ over a given interval $(a \leq x \leq b)$ of interest by dividing this interval into subintervals or "steps." $y(x)$ can be approximated at $(n + 1)$ evenly spaced values of x $(x_0 = a, x_1 = a + h, x_2 = a + 2h, \ldots, x_n = b)$ in accordance with

$$h = \frac{b - a}{n} \qquad (4.65)$$

$$\begin{aligned} x_i &= x_0 + ih \\ &= a + ih \end{aligned} \quad (i = 0, 1, 2, \ldots, n) \qquad (4.66)$$

where h is the chosen step size for the approximation. The difference between the *actual* value $y(x_i)$ of the solution at x_i and the value y_i of the numerical *approximation* at x_i is known as the *discretization error*:

$$\epsilon_i \equiv y_i - y(x_i) \qquad (4.67)$$

(where it is assumed that error due to round-off is not introduced in the numerical calculations).

Taylor's expansion.   Equation (2.54) allows us to obtain an expression for $y(x)$ about a given operating point x_0 via Taylor's series:

$$
\begin{aligned}
y(x) &= y(x_0 + h) \\
&= y(x_0) + hf(x_0, y(x_0)) + \frac{h^2}{2!}f'(x_0, y(x_0)) + \cdots
\end{aligned}
\tag{4.68}
$$

where

$$
\left.\frac{dy}{dx}\right|_{x_0,y_0} = f(x_0, y(x_0)); \qquad \left.\frac{d^2y}{dx^2}\right|_{x_0,y_0} = f'(x_0, y(x_0))
$$

The higher-order derivatives of $f(x, y)$ are obtained via the chain rule (since f is a function of both x and y):

$$
\frac{df}{dx} = \frac{\partial f}{\partial x} + \frac{\partial f}{\partial y}\frac{\partial y}{\partial x}
\tag{4.69}
$$

However, most functions of interest are such that these higher-order derivatives are not easily obtainable. Therefore, the Taylor's series approach is used infrequently except for the simplest case:

$$
y(x_0 + h) \simeq y(x_0) + hf(x_0, y(x_0))
\tag{4.70}
$$

or, in iterative notation,

$$
y_{i+1} = y_i + hf_i \ (i \geq 1)
\tag{4.71a}
$$

$$
y_1 = y(x_0) + hf_0
\tag{4.71b}
$$

$$
f_i \equiv \frac{dy_i}{dx_i}
\tag{4.71c}
$$

[Note that all y_{i+1}, for $i \geq 1$, are based on the value of the previous approximation y_i; y_1 is based on the known (actual) value $y(x_0)$.] Equations (4.71) form the basis for the approximation procedure known as *Euler's method*.

Runge–Kutta methods.   A set of numerical algorithms that involve only first-order derivatives *yet* produce results which are equivalent in accuracy to the higher-order Taylor formulas are known as the Runge–Kutta methods. The general form of the Runge–Kutta algorithm is

$$
y_{i+1} = y_i + h\phi(x, y, h)
\tag{4.72}
$$

where ϕ is the increment function (Henrici, 1962). Define ϕ as a weighted average (with weights of a and b to be chosen by the analyst) of two derivative evaluations k_1 and k_2 on the interval $(x_i \leq x \leq x_{i+1})$:

$$
\phi \equiv ak_1 + bk_2
\tag{4.73}
$$

where

$$k_1 \equiv f(x_i, y_i) \tag{4.74a}$$

$$k_2 \equiv f(x_i + ph, y_i + qhk_1) \tag{4.74b}$$

where p and q are constants that must be determined. Therefore,

$$y_{i+1} = y_i + h(ak_1 + bk_2) \tag{4.75}$$

that is, y_{i+1} is equal to y_i plus the weighted average of the slopes of the function at the points (x_i, y_i) and $(x_i + ph, y_i + qhk_1)$.

It can be shown (see Carnahan et al., 1969) that the four constants a, b, p, and q obey the three relations

$$a = 1 - b \tag{4.76a}$$

$$p = q \tag{4.76b}$$

$$q = \frac{1}{2b} \tag{4.76c}$$

This system is underdetermined (i.e., the number of equations is less than the number of parameters) so that we may arbitrarily choose the value of one variable.

Equation (4.73) is the basis for the *second-order Runge–Kutta method*; in addition to this method, there are higher-order (third-order, fourth-order, etc.) forms of the Runge–Kutta technique which can be used to develop approximate values for y. The development of this higher-order technique is similar to that of the second-order form.

We will consider two special cases of the second-order Runge–Kutta method: one in which the constant b is chosen to be equal to $\frac{1}{2}$ and one in which b is set equal to unity.

1. *Improved Euler's method / Heun's method*: If

$$b \equiv \tfrac{1}{2} \tag{4.77a}$$

then, according to equations (4.76),

$$a = \tfrac{1}{2}, \qquad p = 1, \qquad q = 1 \tag{4.77b}$$

As a result,

$$\bar{y}_{i+1} = y_i + hf(x_i, y_i) \tag{4.78}$$

gives a first approximation (or "prediction") $\bar{y}_{i+1}$ for y_{i+1}, which can then be "corrected" by applying equations (4.73) through (4.76):

$$y_{i+1} = y_i + h(ak_1 + bk_2)$$

or

$$y_{i+1} = y_i + \frac{h}{2}f(x_i, y_i) + f(x_i + h, y_i + hf(x_i, y_i))$$

$$= y_i + \frac{h}{2}f(x_i, y_i) + f(x_{i+1}, \bar{y}_{i+1}) \tag{4.79}$$

Notice that the last term of the expression (4.79) is the weighted average of the slopes $f(x_i, y_i)$ and $f(x_{i+1}, \bar{y}_{i+1})$ at the two ends of the subinterval $x_i \leq x \leq x_{i+1}$. Equations (4.78) and (4.79) form the basis of the simplest "predictor-corrector" method.

2. *Modified Euler's method/improved polygon method*: If

$$b \equiv 1 \tag{4.80a}$$

then, according to equations (4.76),

$$a = 0, \qquad p = \tfrac{1}{2}, \qquad q = \tfrac{1}{2} \tag{4.80b}$$

so that

$$\bar{y}_{i+(1/2)} = y_i + \frac{h}{2}f(x_1, y_i) \tag{4.81}$$

$$y_{i+1} = y_i + hf(x_{i+(1/2)}, \bar{y}_{i+(1/2)}) \tag{4.82}$$

A first approximation $\bar{y}_{i+(1/2)}$ is obtained at $(x_i + h/2)$, that is, at the center of the subinterval $x_i \leq x \leq (x_i + h)$; the slope $f(x, y)$ is then approximated at this position and used to "correct" the initial prediction $\bar{y}_{i+(1/2)}$ in accordance with equation (4.82).

Example 4.5

(Based on Carnahan et al., 1969):   Consider the equation

$$F\left(x, y, \frac{dy}{dx}\right) = 0 = \frac{dy}{dx} - x - y \tag{4.83}$$

or

$$f(x, y) = \frac{dy}{dx}$$

$$= x + y \tag{4.84}$$

$$= 0$$

which we would like to evaluate in the interval $(0 \leq x \leq z)$, We know that the analytical solution $y(x)$ can be obtained as follows for this simple first-order equation (4.84):

$$(D - 1)y = x$$

→   root $r = 1$

→ $y_{ts} = Ce^x = y_{cf}$ = complementary function
                                                   or transient solution

Also,

$$y_{ss} = Ax + B = y_{PI} = \text{particular integral}$$
                                           or steady-state solution

→ $y'_{ss} = A$

→ $(D - 1)y_{ss} = A - Ax - B$

so that

$$A = -1$$

$$B = -1$$

or

$$y = y_{ts} + y_{ss}$$
$$= Ce^x - x - 1$$

If, at $x = 0$,

$$y(0) = 0$$

then

$$C = 1$$

or, finally

$$y = e^x - x - 1 \qquad\qquad (4.85)$$

Table 4.4a presents a program listing in which the second-order Runge–Kutta approximation is used for the special case $(b \equiv \frac{1}{2})$, that is, the improved Euler's method or Heun's method, to evaluate $f(x, y)$ numerically. Sample input and output data are presented in Table 4.4b. Note that as the step size h is decreased, the error ϵ also decreases.

The Runge–Kutta methods (together with any necessary changes of variables) allow us to evaluate numerically many nonlinear ordinary differential equations which describe the behavior of physical systems.

One final comment: There are numerous variations of the techniques described in this chapter. In addition, there are a large number of other numerical approximation methods which can be found in the literature. We will continue to focus on systems with behavior that can be assumed to be

TABLE 4.4a   Sample Computer Program Using the Improved Euler's Method or Huen's Method to Evaluate $f(x, y)$

```
C      PROGRAM:   RUNGE-KUTTA APPROXIMATION
C
C      ...A PROGRAM TO NUMERICALLY ANALYZE THE FUNCTION:
C
C          F(X,Y) = X + Y
C
C      WHERE F(X,Y) IS THE FIRST DERIVATIVE OF Y WITH RESPECT TO X,
C      USING A SECOND-ORDER RUNGE-KUTTA APPROXIMATION (I.E., AN
C      IMPROVED EULER'S METHOD OR HEUN'S METHOD).
C
       IMPLICIT REAL*8 (A-H,O-Z)
C
       OPEN(6,FILE='OUT.TXT')
C
       WRITE(*,*) '  UPPER BOUND: XMAX? '
       READ(*,*)  XMAX
       WRITE(*,*) '  STEP SIZE: H? '
       READ(*,*)  H
       WRITE(*,*) '  PRINT EVERY Kth OUTPUT: K? '
       READ(*,*)  K
C
C      INITIALIZE ZERO CONDITIONS
C
       X      = 0.
       Y      = 0.
       SLOPE  = 0.
       ACTUAL = 0.
       ERROR  = 0.
C
       WRITE(6,100)  XMAX, H, K
       WRITE(6,200)  X, Y, SLOPE, ACTUAL, ERROR
C
       NSTEPS = (XMAX + H/2.) / H
C
       DO 10 I = 1, NSTEPS
          YB     = Y
          XB     = X
          Y      = YB + H * (XB + YB)
          X      = FLOAT(I) * H
          Y      = YB + (H/2.) * (XB + YB + X + Y)
          SLOPE  = X + Y
          ACTUAL = DEXP(X) - X - 1.
          ERROR  = Y - ACTUAL
C
          IF (I/K*K .EQ. I) WRITE(6,200) X,Y,SLOPE,ACTUAL,ERROR
C
   10 CONTINUE
C
       HSQ = H**2
       HCU = H**3
       WRITE(6,300) ERROR, H, HSQ, HCU
C
       CLOSE(6,STATUS='KEEP')
C
  100 FORMAT(//,T4,'2ND-ORDER RUNGE-KUTTA...',//,T4,'UPPER BOUND:',
     1          ' XMAX = ',F4.1,'    STEP SIZE: H = ',F5.3,/,T4,'PRINT',
     2          ' EVERY Kth OUTPUT: K = ',I2,//,T6,'X',10X,'Y',8X,
     3          'SLOPE',5X,'ACTUAL Y',4X,'ERROR',/)
  200 FORMAT(T4,F6.4,4X,F7.5,4X,F7.5,4X,F7.5,4X,F8.6)
  300 FORMAT(/,T4,'ERROR = ',F9.6,'  H = ',F9.6,'  H2 = ',F9.6,
     1          '  H3 = ',F9.6)
C
       END
```

TABLE 4.4b   Sample Input and Output Data for Program Given in Table 4.4a

```
2ND-ORDER RUNGE-KUTTA...

UPPER BOUND: XMAX =  2.0   STEP SIZE: H =  .100
PRINT EVERY Kth OUTPUT: K =  1

   X        Y        SLOPE     ACTUAL Y      ERROR

 .0000    .00000    .00000     .00000      .000000
 .1000    .00500    .10500     .00517     -.000171
 .2000    .02103    .22103     .02140     -.000378
 .3000    .04923    .34923     .04986     -.000626
 .4000    .09090    .49090     .09182     -.000923
 .5000    .14745    .64745     .14872     -.001275
 .6000    .22043    .82043     .22212     -.001690
 .7000    .31157   1.01157     .31375     -.002179
 .8000    .42279   1.22279     .42554     -.002752
 .9000    .55618   1.45618     .55960     -.003421
1.0000    .71408   1.71408     .71828     -.004201
1.1000    .89906   1.99906     .90417     -.005107
1.2000   1.11396   2.31396    1.12012     -.006156
1.3000   1.36193   2.66193    1.36930     -.007370
1.4000   1.64643   3.04643    1.65520     -.008771
1.5000   1.97130   3.47130    1.98169     -.010385
1.6000   2.34079   3.94079    2.35303     -.012242
1.7000   2.75957   4.45957    2.77395     -.014374
1.8000   3.23283   5.03283    3.24965     -.016819
1.9000   3.76628   5.66628    3.78589     -.019619
2.0000   4.36623   6.36623    4.38906     -.022821

ERROR =  -.022821  H =   .100000  H2 =   .010000  H3 =   .001000


2ND-ORDER RUNGE-KUTTA...

UPPER BOUND: XMAX =  2.0   STEP SIZE: H =  .050
PRINT EVERY Kth OUTPUT: K =  2

   X        Y        SLOPE     ACTUAL Y      ERROR

 .0000    .00000    .00000     .00000      .000000
 .1000    .00513    .10513     .00517     -.000044
 .2000    .02130    .22130     .02140     -.000098
 .3000    .04970    .34970     .04986     -.000163
 .4000    .09159    .49159     .09182     -.000239
 .5000    .14839    .64839     .14872     -.000331
 .6000    .22168    .82168     .22212     -.000439
 .7000    .31319   1.01319     .31375     -.000566
 .8000    .42483   1.22483     .42554     -.000714
 .9000    .55871   1.45871     .55960     -.000888
1.0000    .71719   1.71719     .71828     -.001091
1.1000    .90284   2.00284     .90417     -.001326
1.2000   1.11852   2.31852    1.12012     -.001599
1.3000   1.36738   2.66738    1.36930     -.001914
1.4000   1.65292   3.05292    1.65520     -.002278
1.5000   1.97899   3.47899    1.98169     -.002697
1.6000   2.34985   3.94985    2.35303     -.003180
1.7000   2.77021   4.47021    2.77395     -.003734
1.8000   3.24528   5.04528    3.24965     -.004369
1.9000   3.78080   5.68080    3.78589     -.005097
2.0000   4.38313   6.38313    4.38906     -.005929

ERROR =  -.005929  H =   .050000  H2 =   .002500  H3 =   .000125
```

TABLE 4.4b   Sample Input and Output Data (continued)

2ND-ORDER RUNGE-KUTTA...

UPPER BOUND: XMAX = 2.0   STEP SIZE: H = .025
PRINT EVERY Kth OUTPUT: K = 4

X	Y	SLOPE	ACTUAL Y	ERROR
.0000	.00000	.00000	.00000	.000000
.1000	.00516	.10516	.00517	-.000011
.2000	.02138	.22138	.02140	-.000025
.3000	.04982	.34982	.04986	-.000041
.4000	.09176	.49176	.09182	-.000061
.5000	.14864	.64864	.14872	-.000084
.6000	.22201	.82201	.22212	-.000112
.7000	.31361	1.01361	.31375	-.000144
.8000	.42536	1.22536	.42554	-.000182
.9000	.55938	1.45938	.55960	-.000226
1.0000	.71800	1.71800	.71828	-.000278
1.1000	.90383	2.00383	.90417	-.000338
1.2000	1.11971	2.31971	1.12012	-.000407
1.3000	1.36881	2.66881	1.36930	-.000488
1.4000	1.65462	3.05462	1.65520	-.000580
1.5000	1.98100	3.48100	1.98169	-.000687
1.6000	2.35222	3.95222	2.35303	-.000810
1.7000	2.77300	4.47300	2.77395	-.000951
1.8000	3.24853	5.04853	3.24965	-.001113
1.9000	3.78460	5.68460	3.78589	-.001299
2.0000	4.38755	6.38755	4.38906	-.001511

ERROR = -.001511   H =   .025000   H2 =   .000625   H3 =   .000016

2ND-ORDER RUNGE-KUTTA...

UPPER BOUND: XMAX = 2.0   STEP SIZE: H = .010
PRINT EVERY Kth OUTPUT: K = 8

X	Y	SLOPE	ACTUAL Y	ERROR
.0000	.00000	.00000	.00000	.000000
.0800	.00329	.08329	.00329	-.000001
.1600	.01351	.17351	.01351	-.000003
.2400	.03124	.27124	.03125	-.000005
.3200	.05712	.37712	.05713	-.000007
.4000	.09181	.49181	.09182	-.000010
.4800	.13606	.61606	.13607	-.000013
.5600	.19066	.75066	.19067	-.000016
.6400	.25646	.89646	.25648	-.000020
.7200	.33441	1.05441	.33443	-.000024
.8000	.42551	1.22551	.42554	-.000029
.8800	.53086	1.41086	.53090	-.000035
.9600	.65165	1.61165	.65170	-.000041
1.0400	.78917	1.82917	.78922	-.000049
1.1200	.94480	2.06480	.94485	-.000057
1.2000	1.12005	2.32005	1.12012	-.000066
1.2800	1.31656	2.59656	1.31664	-.000076
1.3600	1.53611	2.89611	1.53619	-.000088
1.4400	1.78060	3.22060	1.78070	-.000101
1.5200	2.05211	3.57211	2.05223	-.000115
1.6000	2.35290	3.95290	2.35303	-.000131
1.6800	2.68541	4.36541	2.68556	-.000149
1.7600	3.05227	4.81227	3.05244	-.000169
1.8400	3.45635	5.29635	3.45654	-.000192
1.9200	3.90074	5.82074	3.90096	-.000217
2.0000	4.38881	6.38881	4.38906	-.000244

ERROR = -.000244   H =   .010000   H2 =   .000100   H3 =   .000001

linear over the interval of interest; the simplifications in the analysis of such systems which can be obtained through the application of Laplace transformations are very attractive. However, the reader *should* become familiar with the numerical methods that can be used for the evaluation of nonlinear behavior as the need arises.

4.8 REVIEW

In summary, we have reviewed the following topics, facts, relationships, or concepts in this chapter.

- Several numerical iteration methods have been presented which allow one to obtain an approximate value for a root of a given function $f(x)$. In addition to the familiar quadratic formula, these methods include:

  (a) *Newton–Raphson approximation*, in which the iterative relation is

$$x_{k+1} = x_k - \frac{f(x_k)}{f'(x_k)} \qquad (4.11)$$

  Convergence is not achieved for some functions.

  (b) *Method of successive approximations*, in which the iterative relation is

$$x_{k+1} = h(x_k) \qquad (4.20)$$

  Convergence is guaranteed if $|h'(x_k)| < 1$.

  (c) *Secant method*, for which the iterative relation is

$$x_{k+1} = \frac{x_{k-1}f(x_k) - x_k f(x_{k-1})}{f(x_k) - f(x_{k-1})} \qquad (4.26)$$

  Convergence may not be achieved.

(d) *Regula falsi method* or the *method of false positions*, in which the iterative relation is

$$x_{k+1} = \frac{x_R f(x_L) - x_L f(x_R)}{f(x_L) - f(x_R)}$$ (4.27)

• *Lin's method* allows us to find quadratic factors $(x^2 + px + q)$ of a polynomial

$$p_n(x) = a_n x^n + a_{n-1} x^{n-1} + \cdots + a_2 x^2 + a_1 x + a_0$$
$$= (x^2 + px + q)(b_n x^{n-2} + b_{n-1} x^{n-3} + \cdots + b_3 x^2 + b_2)$$
$$+ b_1 x + b_0$$

by means of repeated application of the relations

$$
\begin{aligned}
a_n &= b_n \\
a_{n-1} &= b_{n-1} + b_n p \\
a_{n-2} &= b_{n-2} + b_{n-1} p + b_n q \\
&\vdots \qquad \vdots \\
a_2 &= b_2 + b_3 p + b_4 q \\
a_1 &= b_2 p + b_3 q \\
a_0 &= b_2 q \\
b_1 &= 0 \\
b_0 &= 0
\end{aligned}
$$ (4.33), (4.35)

The quadratic formula can then be used to determine the roots of the polynomial.

• Ordinary differential equations of order n may be expressed as n first-order equations through a simple change of variables; $(n-1)$ new variables must be introduced.

• Nonlinear and linear first-order ordinary differential equations of the form

$$f(x, y) = \frac{dy}{dx}$$

may be evaluated via

(a) *Euler's method*, for which

$$y_{i+1} = y_i + hf_i$$

$$f_i = \frac{dy_i}{dx_i}$$

$$y_1 = y(x_0) + hf_0$$

(b) *Runge–Kutta methods*, for which

$$y_{i+1} = y_i + h\phi(x, y_i h)$$

where ϕ is the increment function. In particular, we may consider the special cases in which

$$\phi = ak_1 + ak_2$$

$$k_1 = f(x_i, y_i)$$

$$k_2 = f(x_i + ph, y_i + qhk_1)$$

so that

$$a = 1 - b$$

$$p = q$$

$$q = \frac{1}{2b}$$

(1) *Improved Euler's method / Heun's method*:

$$b = \tfrac{1}{2}, \qquad a = \tfrac{1}{2}, \qquad p = 1, \qquad q = 1$$

Thus

$$\bar{y}_1 \equiv y_i + hf(x_i, y_i)$$

$$y_{i+1} = y_i + \frac{h}{2}[f(x_i, y_i) + f(x_{i+1}, \bar{y}_i)]$$

(This is a predictor-corrector method.)

(2) *Modified Euler's method / improved polygon method*:

$$b \equiv 1, \qquad a = 0, \qquad p = \tfrac{1}{2}, \qquad q = \tfrac{1}{2}$$

Thus

$$\bar{y}_{i+(1/2)} = y_i + \frac{h}{2}f(x_i, y_i)$$

$$y_{i+1} = y_i + hf\left(x_i + \frac{h}{2}, \bar{y}_{i+(1/2)}\right)$$

EXERCISES

4.1. Explain the need for numerical approximation methods in systems analysis.

4.2. Consider the function

$$f(x) = 5e^{-2x} + 3$$

Use the Newton-Raphson method to determine the root of $f(x)$ [i.e., the value of x for which $f(x)$ equals zero]. Begin with the initial trial value $x_0 = 0$. Perform five iterations and present your results in a table.

4.3. Consider the function

$$f(x) = 2x + 3 \sin 2x$$

(a) Use the Newton-Raphson method to determine the root of $f(x)$. Begin with the initial trial value $x_0 = 1$. Perform three iterations and present your results in a table.

(b) Draw a graph of $f(x)$ as a function of x. Choose an appropriate range of variation for x so that the root that you obtained in part (a) is shown clearly on the graph.

4.4. Consider the polynomial

$$3x^2 + 2x + 4 = 0$$

Use the method of successive approximations to determine the roots of this polynomial. Perform three iterations.

4.5. Consider the polynomial

$$5x^3 + 2x^2 + 6x + 2 = 0$$

Use the method of successive approximations to determine the roots of this polynomial. Perform three iterations.

4.6. Consider the function

$$x^2 + 3x - 10 = 0$$

(a) Use the Newton-Raphson method to determine the real roots. Use $x_0 = 0$.

(b) Use the method of successive approximations to determine the real roots. Perform three iterations.

4.7. Consider the function

$$x^3 + 4x^2 - 11x - 30 = 0$$

Use the method of false positions (regula falsi) to determine the roots of this polynomial.

4.8. Use Lin's method to determine the roots of the equation

$$x^5 - 0.2x^4 - 5x^3 + x^2 + 4x - 0.8 = 0$$

Perform 10 iterations. You may develop a computer program that will perform this analysis, if you wish.

4.9. Use Lin's method to determine the quadratic factors of the polynomial

$$x^4 + 2x^3 - 19x^2 - 8x + 60 = 0$$

Develop a computer program that will perform this analysis. Express the results in tabular form (for all iterations).

4.10. Use Lin's method to determine the values of v which satisfy equation (4.51) for:
  (a) Air
  (b) Benzene
  (c) Helium
  (d) Water
  in the pressure range 1 to 10 atm. (in 1-atm increments) and for the temperature range 10 to 100°K (in 10°K increments). Write a computer program that will perform this analysis and produce results in tabular form.

4.11. Use the ideal gas equation of state [expression (4.50)] to perform an analysis of the specific volume v for:
  (a) Air
  (b) Benzene
  (c) Helium
  (d) Water
  in the pressure range and temperature range specified in Exercise 4.10. Write a computer program that will perform this analysis and produce results in tabular form.

4.12. Use Lin's method to determine the values of v that satisfy equation (4.51) *and* use the ideal gas equation (4.50) of state to determine these values of v for:
  (a) Acetylene
  (b) Ethylene
  (c) Carbon dioxide
  (d) Helium
  in the pressure range 1 to 10 atm. (in 1-atm increments) and for the temperature range 10 to 100°K (in 10°K increments). Write a program that will perform each of these analyses and will compare the results at each set of (P, T) values. Results should be produced in tabular form. Comment on these results; which mathematical model (the ideal gas equation of state or van der Waals equation) is more accurate (1) at low pressure? (2) at high pressure? How does temperature variation affect your results for each model?

4.13. Consider the function $f(x, y)$ given by the expression

$$f(x, y) = \frac{dy}{dx}$$

$$= 2x + 3y$$

$$= 0$$

  (a) Evaluate $f(x, y)$ analytically in order to determine $y(x)$. Assume that $y(x) = 0$ at $x = 0$.
  (b) Use the improved Euler's method or Huen's method to evaluate $f(x, y)$ for the range $0 \le x \le 5$. Use a step size h equal to 0.1.

(c) Use the modified Euler's method or the improved polygon method to evaluate $f(x, y)$ for the range $0 \le x \le 5$. Use a step size h equal to 0.1.

(d) Compare the results of parts (a), (b), and (c). Comment on the relative accuracy of the improved Euler's method and the modified Euler's method for this example.

(e) Use the improved Euler's method to evaluate $f(x, y)$ with h equal to 0.01. Also, perform this evaluation with a step size h equal to 0.001. Compare and comment on the relative accuracy of these two evaluations.

4.14 Use the approximation method of your choice to evaluate the equation

$$\frac{dy}{dx} + y - 4x^2 - 3x - 2 = 0$$

(a) for $y(x)$, where $y(x)$ is equal to zero at $x = 0$. Perform your evaluation over the interval $x = 0$ to $x = 10$. Compare your numerical results with the analytical solution.

(b) for $y(x)$, where $y(x)$ is equal to 2 at $x = 0$. Perform your evaluation over the interval $x = 0$ to $x = 100$. Compare your numerical results with the analytical solution.

4.15. Evaluate the equation given in Exercise 4.14 by using the improved Euler method or Huen's method for the cases where the step size h is given by

(a) $h = 0.1$

(b) $h = 0.01$

(c) $h = 0.001$

Compare the error ϵ obtained in each case and comment on the effect of the step size on this error. Use the interval $x = 0$ to $x = 10$ for your evaluations.

4.16. Evaluate the equation

$$f(x, y) = \frac{dy}{dx} = y + 4x + 5e^{-x} = 0$$

for $y(x)$, with zero initial conditions, over the interval $x = 0$ to $x = 3$. Use the following methods and compare your results with the analytical solution.

(a) Improved Euler's method

(b) Modified Euler's method

Select an appropriate step size h for each case.

4.17 Use Lin's method to determine the roots of the characteristic equation

$$3s^5 + 4s^4 + 2s^3 + 18s^2 + s + 16 = 0$$

Perform 10 iterations by writing a computer program that will evaluate the expression. The program should also provide the user with a statement regarding the stability of the system described by the equation.

4.18. Consider the following two simultaneous differential equations for a system:

$$\frac{d^2y_1}{dt^2} + \frac{dy_1}{dt} + y_1 + \frac{dy_2}{dt} + y_2 = 3 \sin 10t$$

$$\frac{dy_1}{dt} + \frac{d^2y_2}{dt^2} + 2\frac{dy_2}{dt} + y_2 = 3t$$

(a) Obtain a single differential equation in which only y_2 appears as an unknown dependent variable.

(b) Determine $y_2(t)$ under the assumption of initial zero conditions. Use any appropriate numerical technique(s) that is necessary to perform this evaluation.

4.19. Consider the characteristic equation

$$s^3 + 2s^2 + 16 = 0$$

Use the method of false positions (regula falsi) to determine the roots of this equation.

4.20 Consider the characteristic equation

$$s^5 + 4s^4 + 3s^3 + 17s^2 + 9s + 12 = 0$$

Use Lin's method to determine the quadratic and linear factors of this equation. Develop a computer program that will perform such an evaluation and present the results in an appropriate form.

4.21. Consider the following mechanical system:

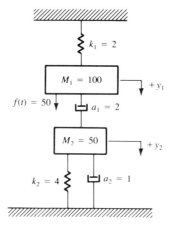

The corresponding set of second-order differential equations which describe the behavior of this system are

$$M_1 \frac{d^2 y_1}{dt^2} + a_1 \frac{dy_1}{dt} + K_1 y_1 - a_1 \frac{dy_2}{dt} = f(t)$$

$$M_2 \frac{d^2 y_2}{dt^2} + (a_1 + a_2) \frac{dy_2}{dt} + K_2 y_2 - a_1 \frac{dy_1}{dt} = 0$$

(a) Obtain a single fourth-order differential equation in which only $y_1(t)$ appears as a dependent variable.

(b) Assuming zero initial conditions, evaluate $y_1(t)$. You may use any numerical approximation technique that you deem to be necessary and appropriate. Present your results for $x(t)$ as a function of time in tabular form.

4.22 Determine the displacement $x_1(t)$ of the mass M_1 in the system shown in Exercise 2.19. Assume zero initial conditions and system parameters given by

$$a = 3, \quad k_1 = k_2 = 2, \quad k_3 = 4, \quad M_1 = 10, \quad M_2 = 8, \quad f(t) = 25$$

4.23. Evaluate the roots of the characteristic equation

$$s^6 + 3s^5 + 2s^4 + s^3 + 2s^2 + s + 50 = 0$$

Use any numerical approximation technique that you deem to be appropriate.

4.24. Evaluate the roots of the characteristic equation

$$s^4 + 4s^3 + 2s^2 + s + 25 = 0$$

Use any numerical approximation technique that you deem to be appropriate.

4.25. Evaluate the roots of the characteristic equation

$$s^3 + 3s^2 + 2s + 15 = 0$$

Use any numerical approximation technique that you deem to be appropriate.

4.26. Evaluate the roots of the characteristic equation

$$2s^5 + 4s^4 + 3s^3 + 3s^2 + 2s + 5 = 0$$

Use any numerical approximation technique that you deem to be appropriate.

5

TRANSFER FUNCTIONS AND BLOCK DIAGRAMS

*You have erred perhaps in attempting to put colour and life into each of
your statements instead of confining yourself to the task of placing
upon record that severe reasoning from cause to effect which is really the
only notable feature about the thing.*

Sherlock Holmes, The Adventure of the Copper Beechers
by Sir Arthur Conan Doyle

5.1 OBJECTIVES

Upon completion of this chapter, the reader should be able to:

- Determine the transfer function for a particular system or subsystem.
- Apply block diagram notation to describe the structure or organization of a system and the flow of information (signals) through that system.
- Manipulate, by means of block diagram algebra, the block diagram description of a system into equivalent (and perhaps more desirable or useful) forms.

5.2 TRANSFER FUNCTIONS

Systems can be very effectively (and simply) represented by *block diagrams*; such diagrams allow us to express the similarities between various types (electrical, mechanical, thermal, etc.) of systems in a form that details the organizational structure of the system together with the variables linking system components.

The general form of a block diagram component is shown in Figure 5.1, where $I(s)$ is the input (forcing function) to the system, $O(s)$ denotes the system output, and T represents the transfer function for this component. The *transfer function* represents the effect of the component on the input I in order

Figure 5.1 Block diagram representation of a component with transfer function $T(s)$.

to produce the output O; it is defined in the Laplace domain as the ratio of the Laplace transform $O(s)$ of the output with respect to the Laplace transform $I(s)$ of the input, that is,

$$T(s) \equiv \frac{O(s)}{I(s)} \tag{5.1}$$

This relation is valid under the assumption that *all initial conditions are zero*; it applies only to linear time-invariant systems [although its application can be extended to certain nonlinear systems; see Ogata (1970)].

The transfer function $T(s)$ obeys the principle of superposition, that is, for multiple forcing functions $I_1(s)$, $I_2(s)$, ... —which are equivalent to a total input $I(s)$ given by

$$I(s) = a_1 I_1(s) + a_2 I_2(s) + \cdots \tag{5.2}$$

(where $a_1, a_2, \ldots$ are appropriate constant coefficients)—the response $O(s)$ for the system is given by

$$
\begin{aligned}
O(s) &= T(s)I(s) \\
&= a_1 T(s)I_1(s) + a_2 T(s)I_2(s) + \cdots \\
&= a_1 O_1(s) + a_2 O_2(s) + \cdots
\end{aligned} \tag{5.3}
$$

The transfer function $T(s)$ is a rational function of s with coefficients that depend only on the elements of the system. Therefore (as described graphically in Figure 5.1), $T(s)$ is a *property of the system* (or component); it describes the operation of a component on its input $I(s)$. $T(s)$ is independent of the forcing function $I(s)$.

5.3 BLOCK DIAGRAMS

Block diagrams allow us to indicate clearly the flow of signals (i.e., information) through a system by directed arrows; the functional blocks denote system components (or combinations of components) which operate on these input signals in accordance with the transfer function associated with the block— thereby producing the given output signal.

Different types of systems (e.g., electrical, mechanical, thermal) may behave according to the same type of differential equation relating the input $I(t)$ and the output $O(t)$; as a result, the same block diagram can be used to represent these dissimilar systems. Additionally, a single system can be represented by different types of (equivalent) block diagrams which are chosen in accordance with the needs of the system analyst/designer.

In a block diagram, one does *not* indicate the primary source of energy for the system; this source is external to the system under consideration.

The *error detector* represents a sensing component that compares a reference input $R(s)$ (notice that we are in the Laplace domain in which the transfer function is properly defined) to a feedback signal $B(s)$ in order to produce an actuating signal $E(s)$. Such a comparison may be of the form described mathematically by

$$E(s) = R(s) - B(s) \qquad (5.4)$$

in which one measures error in terms of the difference between $R(s)$ and the system output $C(s)$:

$$\epsilon(s) \equiv R(s) - C(s) \qquad (5.5)$$

(The corrective actuating signal E is proportional to the error ϵ.)

If corrective action is to be initiated by a threshold value of $B(s)$, then

$$E(s) = R(s) + B(s) \qquad (5.6)$$

in which case the detector is known as a *sum operator*; equation (5.4) applies to a *difference operator*. The error detector is denoted in block diagrams by the symbol

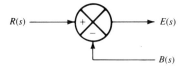

if equation (5.4) describes the relation between the signals $E(s)$, $R(s)$, and $B(s)$; for the case of sum operators, the symbol to be used is

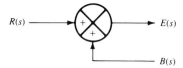

Additional sensing devices can be of the addition/subtraction type:

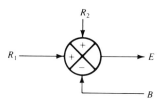

for which

$$E(s) = R_1(s) + R_2(s) - B(s) \tag{5.7}$$

and the multiplication type:

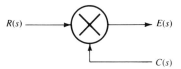

for which

$$E(s) = R(s)C(s) \tag{5.8}$$

Notice that there is only one output signal $E(s)$ from an error detector. Potentiometers, differential amplifiers, synchros, multipliers, and so on, are devices which are used as error detectors (see Kuo, 1975).

We may now represent a closed-loop system by a feedback-loop block diagram (Figure 5.2), where we have used the standard notation in which $G(s)$ denotes the feedforward (forward-loop) component's transfer function and $H(s)$ represents the feedback-loop (measuring) component's transfer function, such that the feedback signal $B(s)$ fed to the error detector is related to the system's output signal $C(s)$ according to the relation

$$B(s) = H(s)C(s) \tag{5.9}$$

The output signal $C(s)$ can be simultaneously directed to multiple sites (blocks, detectors, or other systems) from the branch point denoted as *bp* in Figure 5.2.

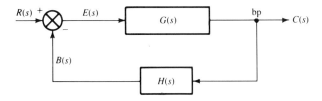

Figure 5.2   Block diagram representation of a closed-loop (feedback) system.

We may reduce the closed-loop block diagram of Figure 5.2 to an *equivalent open-loop diagram* by identifying the transfer function $T(s)$ which allows us to state the equivalency relation between these diagrams as shown in Figure 5.3. In other words, we seek the equivalent transfer function $T(s)$ which, operating upon $R(s)$, produces the output signal $C(s)$. We use the relation

$$E(s) = R(s) - B(s) \tag{5.10}$$

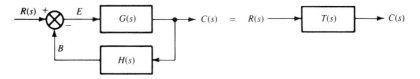

Figure 5.3 Equivalent block diagram representations.

together with equation (5.9) to obtain

$$E(s) = R(s) - H(s)C(s) \qquad (5.11)$$

Then we note that [after (at least) one loop has been completed for the system's signals]

$$C(s) = G(s)E(s) \qquad (5.12)$$

Equation (5.11) then allows us to write the following first-order approximation for expression (5.12):

$$\begin{aligned} C(s) &= G(s)[R(s) - H(s)C(s)] \\ &= G(s)R(s) - G(s)H(s)C(s) \end{aligned} \qquad (5.13)$$

Solving this expression for $C(s)$, we have

$$C(s) = \frac{G(s)}{1 + G(s)H(s)} R(s) \qquad (5.14)$$

or, equivalently,

$$C(s) = T(s)R(s) \qquad (5.15)$$

where $T(s)$ is the equivalent transfer function which we have sought:

$$T(s) \equiv \frac{G(s)}{1 + G(s)H(s)} \qquad (5.16)$$

In other words, a closed-loop diagram can be expressed as the equivalent open-loop diagram with an (equivalent) transfer function $T(s)$ given by expression (5.16).

5.4 BLOCK DIAGRAM ALGEBRA AND REDUCTION

A set of system components, acting in series, can be represented by a block diagram of the form shown in Figure 5.4a; the relations between signals can

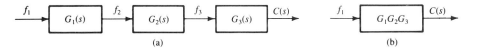

Figure 5.4   Equivalent representations.

be represented by the expressions

$$f_2 = G_1 f_1 \tag{5.17a}$$

$$f_3 = G_2 f_2 \tag{5.17b}$$

$$C = G_3 f_3 \tag{5.17c}$$

or

$$\begin{aligned} C &= G_3(G_2 f_2) \\ &= G_3 G_2(G_1 f_1) \\ &= G_3 G_2 G_1 f_1 \\ &= G_1 G_2 G_3 f_1 \end{aligned} \tag{5.18}$$

(where we have intentionally chosen to simplify our notation by not indicating explicitly that each of these signals and transfer functions are evaluated in the Laplace domain; that is, they are functions of s). As a result, a reduced form of this block diagram is shown in Figure 5.4b.

Such block diagram "algebra" allows us to reduce a complex system structure to an (equivalent) simple diagram, although the equivalent transfer function for the reduced diagram will be more complex than those of the initial diagram's functional blocks. Reduction of block diagrams increases our ability to recognize analogous systems.

As another example, consider the block diagram of Figure 5.5a; signals are related in accordance with the following relations:

$$f_2 = G_1 f_1 \tag{5.19a}$$

$$f_4 = G_2 f_3 \tag{5.19b}$$

$$\begin{aligned} E &= f_2 - f_4 \\ &= G_1 f_1 - G_2 f_3 \end{aligned} \tag{5.19c}$$

An equivalent diagram—which also obeys these relations—is shown in Figure 5.5b. Therefore, we can move a component (with transfer function G_2) from a feedback loop to a feedforward loop by properly modifying another component within the system. As a result, we may develop alternative (yet equivalent) organizational structures for the system. This ability to modify the system's structure extends our opportunity to use different types of system components in a wide variety of possible combinations.

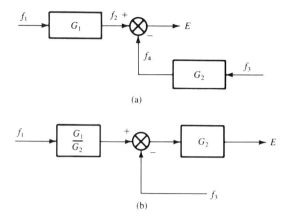

(a)

(b)

Figure 5.5   Manipulating block components to form equivalent representations.

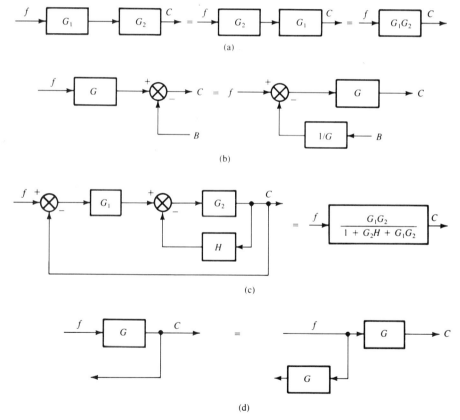

Figure 5.6   Several sets of equivalent block diagram representations for different system structures.

Figure 5.6 presents several sets of equivalent relations between block diagrams.

Appendix D presents a brief review of another graphical format with which one may represent the organizational structure of a physical system; this graphical format is known as *bond graphing*. Bond graphs provide a simple, shorthand notation for differential equations that relate system variables. Such graphs can be easily interpreted by workers with differing technical and mathematical backgrounds, thereby facilitating interdisciplinary discussion and exchange of information. The reader is encouraged to review Appendix D and compare the advantages of bond graphing with those of block diagrams; consider the circumstances in which one would prefer to use bond graphing as an alternative to the block diagram format for systems representation.

5.5 REVIEW

In summary, we have reviewed the following topics, facts, relationships, or concepts in this chapter.

- The transfer function $T(s)$ of a system has been defined as the ratio of the Laplace transform $O(s)$ of the output with respect to the Laplace transform $I(s)$ of the input:

$$T(s) = \frac{O(s)}{I(s)} \tag{5.1}$$

- $T(s)$ is a property of the system; it represents the effect of the system on the input I in order to produce the output O. $T(s)$ is properly defined under the assumption of zero initial conditions.
- The equivalent open-loop transfer function for a closed-loop system is

$$T(s) = \frac{G(s)}{1 + G(s)H(s)} \tag{5.16}$$

where $G(s)$ is the product of the transfer functions of all feedforward-loop components and $H(s)$ is the product of the transfer functions of all feedback-loop components.

- Block diagram notation has been introduced which allows one to represent a system graphically in the form

$$I(s) \longrightarrow \boxed{T(s)} \longrightarrow O(s)$$

- Block diagram notation allows us to represent clearly the similarities between systems with identical transfer functions. In addition, block diagram notation allows us to represent the structure or organization of the system, together with the flow of information (signals) through the system.

EXERCISES

5.1. Determine the transfer function $T(s)$ for the equivalent open-loop block diagram representation

$$I(s) \rightarrow \boxed{T(s)} \longrightarrow O(s)$$

of the following closed-loop system:

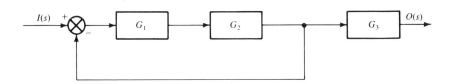

5.2. Determine the equivalent open-loop transfer function $T(s)$ for the following closed-loop system:

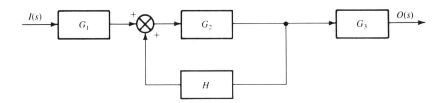

5.3. Consider the following differential equation, which describes the behavior of a physical system:

$$\frac{d^2y}{dt^2} + 2\frac{dy}{dt} + y = 3 \sin 4t + e^{-5t} = f_1(t) + f_2(t)$$

$y(t)$ represents the output or response of the system, given the two forcing functions or inputs $f_1(t)$ and $f_2(t)$.
(a) Determine the Laplace transform of the equation.
(b) Determine the transfer function $T(s)$ for this system.

5.4. Consider the following differential equation, which describes the behavior of a physical system (where D represents the differential operator d/dt):

$$(4D^3 + 3D^2 + 3D + 12)y(t) = 4t + 3e^{-2t} \sin 3t$$

$y(t)$ represent the response of the system. Determine the transfer function $T(s)$ for this system.

5.5. Consider the following block diagram representation of a system:

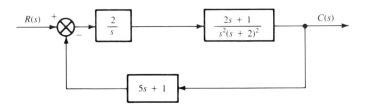

(a) Obtain the equivalent open-loop transfer function $T(s)$ for this system and develop the equivalent open-loop block diagram.
(b) Obtain an equivalent closed-loop block diagram representation for this system with unity feedback.

Problems 5.6 through 5.8 refer to the system shown below.

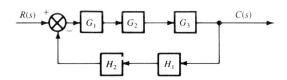

5.6. If

$$G_1 = \frac{2}{s+1}, \quad G_2 = \frac{1}{s^2}, \quad G_3 = \frac{36}{(s^2+2)(s-1)}, \quad H_1 = \frac{4}{s}, \quad H_2 = 2s + 1$$

determine the equivalent open-loop transfer function $T(s)$.

5.7. If $G_1 = G_2 = H_1 = 1$ and $G_3 = 2/s$:

(a) Obtain the equivalent open-loop transfer function $T(s)$ when $H_2 = 2/(s + 3)$.

(b) Obtain the equivalent open-loop transfer function $T(s)$ when

$$H_2 = \frac{1}{s(s + 1)}$$

5.8. Given that

$$G_1 = \frac{4}{s}, \quad G_2 = \frac{48}{2s(s + 3)}, \quad G_3 = 1, \quad H_1 = 40s + 5, \quad H_2 = 1$$

(a) Obtain the equivalent open-loop transfer function $T(s)$.

(b) Obtain the equivalent closed-loop block diagram in which unity feedback is used.

6

SYSTEM ANALYSIS

In five minutes you will say that it is all so absurdly simple.

Sherlock Holmes, The Adventure of the Dancing Men
by Sir Arthur Conan Doyle

6.1 OBJECTIVES

Upon completion of this chapter, the reader should be able to:

- Identify zeroth-order, first-order, and general first-order and second-order systems, together with the corresponding transfer functions for such systems.

- Evaluate the response of zeroth-order, first-order, and second-order systems to various forcing functions, including step impulse, and sinusoidal input functions.

- Recognize and explain the significance of the damping ratio ζ for a system in terms of the expected system behavior. In particular, the reader should be able to determine the value of the damping ratio for a given system, together with the corresponding behavior of the system (i.e., undamped, critically damped, underdamped, or overdamped behavior).

- Explain the concepts of decay ratio, overshoot, rise time, and settling time, together with the significance of the time constant τ for a system.

- Describe the concept of energy domains and analogous systems within these domains.

- Identify "through" variables and "across" variables in each of the energy domains that are reviewed in this chapter, together with the generalized system parameters, which can be categorized as inductive (energy storage), capacitive (energy storage), or resistive (energy dissipative) in nature.

- Write the generalized equations that describe the behavior of inductive, capacitive, and resistive system elements.
- Apply the method of system analogies to evaluate the behavior of equivalent systems in different energy domains.

6.2 ZEROTH-ORDER AND FIRST-ORDER SYSTEMS

We now have the mathematical and graphical tools to design and analyze physical systems. In this chapter we develop transfer functions for zeroth-, first-, and second-order systems, from which more complicated systems can be generated.

Our development of zeroth-, first-, and second-order systems will be performed in the mechanical energy domain, a domain in which most of us have strong intuitive understanding of the behavior of physical components (e.g., springs, dashpots, masses). We will then identify analogies between systems in different energy domains (electrical, mechanical, thermal, etc.), which will allow us to extend our abilities to analyze systems beyond those in the mechanical domain.

6.2.1 Zeroth-Order Systems

Consider the zeroth-order mechanical system shown in Figure 6.1a. A spring (with spring constant k) is attached to a wall and a constant force f is applied to the spring, such that

$$f = \begin{cases} 0 & \text{for } t < 0 \\ \text{constant} & \text{for } t \geq 0 \end{cases} \tag{6.1}$$

(i.e., the forcing function is a step input), thereby resulting in a compression of distance x. The zeroth-order equation that describes the behavior of this

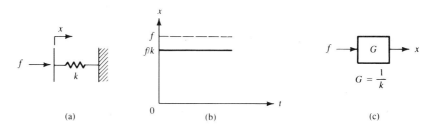

Figure 6.1   Zeroth-order system.

elementary system is

$$kx = f \tag{6.2}$$

or

$$x = \frac{f}{k} \tag{6.3}$$

Figure 6.1b presents the diagram for $x(t)$; the applied force is initiated at time $t = 0$. Both f and x are approximated by a step change. Figure 6.1c shows the block diagram for this system, where f is the input applied to the system and x is the output. The transfer function $G(s)$, defined as the ratio of the Laplace transform of the output with respect to the Laplace transform of the input, is given by

$$
\begin{aligned}
G(s) &\equiv \frac{x(s)}{f(s)} \\
&= \frac{1}{k}
\end{aligned}
\tag{6.4}
$$

under the assumption of zero initial conditions ($x = x' = x'' = \cdots = 0$).

6.2.2 First-Order Systems

A first-order mechanical system is shown in Figure 6.2a: a dashpot (which dissipates energy through viscous damping—due to fluid friction at low and moderate speeds—such that the frictional force is directly proportional to the speed of the system) is attached to a wall with a constant force f applied at time $t \geq 0$. The first-order differential equation for this system relates the velocity v to the force f according to

$$av = a\frac{dx}{dt} = f \tag{6.5}$$

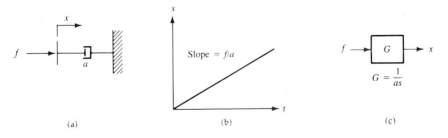

(a)   (b)   (c)

Figure 6.2   First-order system.

where the constant a is known as *the coefficient of viscous damping* (see Beer and Johnston, 1976). In operator notation, this expression becomes

$$aDx = f \tag{6.6}$$

Integration then gives

$$x = \frac{f}{a} \int_0^t dt$$
$$= \frac{f}{a} t + C \tag{6.7}$$

If the dashpot is in equilibrium initially ($x = 0$ at $t = 0$), the constant C of integration must be zero:

$$x = \frac{f}{a} t \tag{6.8}$$

[Equation (6.8) is known as a *ramp* function with a slope equal to f/a and an intercept equal to zero.] The diagram for $x(t)$ is given in Figure 6.2b, with the block diagram for the system shown in Figure 6.2c. The transfer function for this system is

$$G(s) = \frac{x(s)}{f(s)}$$
$$= \frac{1}{as} \tag{6.9}$$

[Notice that, in the time domain,

$$G(t) = \frac{x(t)}{f(t)}$$
$$= \frac{1}{aD} \tag{6.10}$$

although the transfer function is properly defined in the Laplace s-domain. For $f(t)$ equal to a constant, and for zero initial conditions—as always assumed for transfer functions, one simply replaces D in this equation with the Laplace variable s in order to obtain $G(s)$.]

6.3 GENERAL FIRST-ORDER SYSTEMS

6.3.1 Response to an Applied Step Function

A general first-order system includes both zeroth-order and first-order components; for example, the mechanical system shown in Figure 6.3a consists

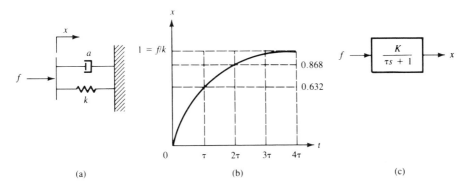

Figure 6.3   General first-order system.

of a dashpot and a spring with a constant applied force f [i.e., f obeys the step-function relationship given in equation (6.1)]. The differential equation for this system reflects the effect of each component:

$$a\frac{dx}{dt} + kx = f \qquad (6.11a)$$

or

$$(aD + k)x = f \qquad (6.11b)$$

The solution to this equation (with a single root r equal to $-k/a$) is given by

$$x(t) = Ce^{-kt/a} + A \qquad (6.12)$$

where A is a constant output (a displacement of the mechanical system) due to the constant input f. The initial condition $(x = 0)$ at $t = 0$ gives

$$A = -C$$

so that

$$x(t) = A(1 - e^{-kt/a}) \qquad (6.13)$$

Furthermore, for zero initial conditions, equation (6.11) becomes

$$a\frac{dx}{dt} = f$$

or, together with equation (6.13),

$$a\left(A\frac{k}{a}e^{-k(0)/a}\right) = Ak(1)$$

or

$$A = \frac{f}{k} \qquad (6.14)$$

Therefore,

$$x(t) = \frac{f}{k}(1 - e^{-kt/a}) \tag{6.15}$$

Figure 6.3b presents $x(t)$ (where f has been normalized such that f/k is equal to unity); if one defines a *time constant* τ for the system such that

$$\tau \equiv \frac{a}{k} \tag{6.16}$$

when $(k/a)t = t/\tau = 1$, $x(t)$ can be evaluated at $t = \tau$, $t = 2\tau$, $t = 3\tau$, and so on. As can be seen in Figure 6.3b, the final steady-state condition (such that $x = f/k$) is *essentially* achieved after a period of operation equal to a time interval of 4τ:

$$x(4\tau) = 0.982 \frac{f}{k} \tag{6.17}$$

so that $x(t)$ does not vary more than 2% from its final steady-state value after $t = 4\tau$.

The block diagram for this general first-order system is shown in Figure 6.3c. The transfer function for the system is given by

$$G(s) = \frac{x(s)}{f(s)}$$

$$= \frac{1}{as + k}$$

$$= \frac{1/k}{(as/k) + 1} \tag{6.18}$$

$$= \frac{1/k}{\tau s + 1}$$

$$= \frac{K}{\tau s + 1}$$

where we have introduced the constant K (known as the "gain" associated with this system), defined by

$$K \equiv \frac{1}{k} \tag{6.19}$$

[In the time domain, one may consider

$$G(t) = \frac{1}{aD + k}$$

$$= \left.\frac{K}{\tau D + 1}\right]$$

(6.20)

Notice that the application of $G(s)$ to $f(s)$ will produce $x(s)$; an inverse Laplace transform will then generate $x(t)$ as given in equation (6.15):

$$x(s) = G(s)f(s)$$

$$= \frac{K/\tau}{s + (1/\tau)}\frac{f}{s}$$

(6.21)

$$x(s) = \frac{C_1}{s + (1/\tau)} + \frac{C_2}{s}$$

where $C_1 = -Kf$ and $C_2 = Kf$. Therefore,

$$x(t) = Kf(1 - e^{-t/\tau})$$

or, for $K = 1/k$ and $\tau = a/k$,

$$x(t) = \frac{f}{k}(1 - e^{-kt/a})$$

(6.15)

Example 6.1

Chin (1984) used a first-order differential equation to model and analyze human productivity:

$$\frac{dp}{dt} + \gamma p(t) = f(t)$$

where $p(t)$ represents human productivity (of any individual or group) and γ denotes a decay constant for $p(t)$ which reflects human inertia within any production effort. The forcing function $f(t)$ represents any concrete form of encouragement (promotion, an increase in wages, etc.) Performing a Laplace transform on this equation, we then obtain

$$(s + \gamma)p(s) = f(s)$$

where we have assumed zero initial conditions; $p(s)$ and $f(s)$ represent the Laplace transforms of $p(t)$ and $f(t)$, respectively. The transfer function for this system is then given by

$$G(s) = \frac{p(s)}{f(s)}$$

$$= \frac{1}{s + \gamma}$$

so that the corresponding block diagram for the system is

$$f(s) \longrightarrow \boxed{\dfrac{1}{s + \gamma}} \longrightarrow p(s)$$

If we assume that $f(t)$ is a step function with a constant value A for $t \geq 0$, we then obtain

$$p(s) = \frac{A}{s(s + \gamma)}$$

where $f(s)$ is equal to A/s. Partial-fraction expansion then produces

$$p(s) = \frac{C_1}{s} + \frac{C_2}{s + \gamma}$$

An inverse Laplace transformation then gives

$$p(t) = C_1 + C_2 e^{-\gamma t}$$

The coefficient C_1 may be evaluated by equating the following terms:

$$\frac{A}{s(s + \gamma)} = \frac{C_1}{s} + \frac{C_2}{s + \gamma}$$

after which the entire expression is multiplied by s:

$$\frac{A}{s + \gamma} = C_1 + \frac{C_2 s}{s + \gamma}$$

Finally, s is set equal to zero, producing the final result:

$$C_1 = \frac{A}{\gamma}$$

Similarly, we may evaluate C_2, which is found to be given as

$$C_2 = -\frac{A}{\gamma}$$

Human productivity is then expected to be approximately described by the expression

$$p(t) = \frac{A}{\gamma}(1 - e^{-\gamma t})$$

As $t \to \infty$, $p(t)$ exponentially approaches a steady-state value equal to A/γ; in other words, as one increases the amount A of encouragement within a production system, he/she can expect that the steady-state level of productivity will increase proportionately. Compare the expression for $p(t)$ above with equation (6.15) for the general first-order mechanical system. Figure 6.3b describes the behavior of both systems, where the decay constant γ for the production process is related to a time constant (or decay time) τ for this system according to

$$\gamma^{-1} = \tau$$

The final steady-state level for $p(t)$ is achieved more swiftly as τ decreases (or, equivalently, as γ increases); recall that the steady-state condition is essentially achieved after a period of operation (i.e., applied encouragement) equal to 4τ [see equation (6.17)]. However, as γ increases, the value A/γ of the steady-state output of the system decreases. As a result, one needs to consider his/her goals for the system in terms of the final productivity *level* which is needed and the *rate* at which this final level is achieved. (Of course, one must also consider if the value of γ within the system can, in fact, be modified or controlled.)

6.3.2 Response to an Applied Impulse Function

For an applied forcing function of the form

$$f(t) = A\delta(t) \tag{6.22a}$$

[where

$$f(t) = \begin{cases} 0 & \text{for } t < 0 \\ \dfrac{A}{\Delta t} & \text{for } 0 \le t \le \Delta t \\ 0 & \text{for } t > \Delta t \end{cases} \tag{6.22b}$$

and

$$\lim_{\Delta t \to 0} f(t) = A \tag{6.23}$$

that is, (t) is the unit-impulse function δ] (see Figure 6.4a), the Laplace transform of $f(t)$ is

$$f(s) = A \tag{6.24}$$

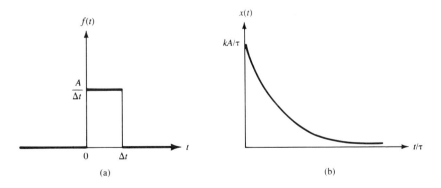

Figure 6.4   Applied impulse function and the response to such a function.

The response $x(t)$ of a first-order system to such an applied impulse function can be obtained as follows:

$$x(s) = G(s)f(s)$$

$$= \left[\frac{K/\tau}{s + (1/\tau)} \right] A$$

(6.25)

which, upon evaluation, produces

$$x(t) = \frac{KA}{\tau} e^{-t/\tau}$$

(6.26)

In other words, the system experiences an abrupt displacement of magnitude KA/τ at time $t = 0$, after which the system exponentially returns to its equilibrium position of $x = 0$. Intuitively, this is the behavior that we would expect from the first-order system of Figure 6.3a upon application of an impulse forcing function.

Example 6.2

Returning to Chin's (1984) general first-order model of human productivity (see Example 6.1), expressed by the equation

$$\frac{dp}{dt} + p(t) = f(t)$$

where $p(t)$ represents human productivity, γ denotes a decay constant for $p(t)$ which reflects human inertia within the system, and $f(t)$ is any concrete form of encouragement (i.e., a forcing function applied to the production system), we assume that $f(t)$ is an impulse function of the form

$$f(t) = A\delta(t)$$

(where A is the height of the impulse); equation (6.24) gives its Laplace transform $f(s)$:

$$f(s) = A$$

A Laplace transformation of the general first-order equation for this system results in the expression

$$(s + \gamma)p(s) = A$$

or

$$p(s) = \frac{A}{s + \gamma}$$

An inverse transformation then produces

$$p(t) = Ae^{-\gamma t}$$

In other words, a short-term application of some form of encouragement (equivalent to an applied impulse) can be expected to produce a corresponding increase in the level of productivity which will then exponentially decay.

6.3.3 Response to an Applied Sinusoidal Function

For an applied forcing function of the form

$$f(t) = A \sin \omega t \tag{6.27}$$

for $t \geq 0$,

$$f(s) = A \left[\frac{\omega}{s^2 + \omega^2} \right] \tag{6.28}$$

The response of a first-order system is then given by (where K has been set equal to unity for simplicity)

$$x(s) = G(s)f(s) \tag{6.29}$$

$$= \frac{1/\tau}{s + (1/\tau)} A \frac{\omega}{s^2 + \omega^2}$$

from which one may determine that

$$x(t) = \frac{A\tau\omega e^{-t/\tau}}{\tau^2\omega^2 + 1} - \frac{A\omega\tau}{\tau^2\omega^2 + 1} \cos \omega t + \frac{A}{\tau^2\omega^2 + 1} \sin \omega t \tag{6.30}$$

If we then apply the identity

$$p \cos \beta + q \sin \beta = R \sin (\beta + \theta) \tag{6.31}$$

where

$$R = (p^2 + q^2)^{1/2} \tag{6.32}$$

and

$$\tan \theta = \frac{p}{q} \tag{6.33}$$

then

$$x(t) = \frac{A\omega\tau}{\tau^2\omega^2 + 1} e^{-t/\tau} + \frac{A}{(\tau^2\omega^2 + 1)^{1/2}} \sin (\omega t + \phi) \tag{6.34}$$

where

$$\phi = \tan^{-1} -\omega\tau \tag{6.35}$$

Therefore, as $t \rightarrow \infty$,

$$x(t) \rightarrow \frac{A}{(\tau^2\omega^2 + 1)^{1/2}} \sin (\omega t + \phi) \tag{6.36}$$

Figure 6.5 shows this response function; *after* the steady-state behavior has been achieved, one observes that:

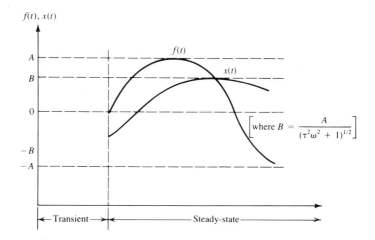

Figure 6.5 Applied sinusoidal function and the response to such a function.

1. The response $x(t)$ oscillates with the frequency ω of the applied forcing function $f(t)$.

2. $x(t)$ has an amplitude which is less than that of the input:

$$\frac{A}{(\tau^2\omega^2 + 1)^{1/2}} < A$$

Such an attenuation of the input signal reflects the expected loss of energy within the system. [If energy is *not* dissipated within the system, then

$$a = 0$$

or

$$\tau = \frac{a}{k}$$

$$= 0$$

so that

$$\frac{A}{(\tau^2\omega^2 + 1)^{1/2}} \rightarrow A$$

that is, the amplitude of the output signal is identical to that of the input signal.]

3. $x(t)$ and $f(t)$ oscillate out of phase due to the phase factor ϕ (reflecting the time necessary for the input signal to be transmitted through the system and a response to be generated by the system).

6.4 GENERAL SECOND-ORDER SYSTEMS

Consider the mechanical system shown in Figure 6.6a. A dashpot (an energy dissipator) and a spring (an energy storage element) are connected to a mass m (another energy storage element); a constant force f is applied to the system for time $t \geq 0$. Newton's second law of motion, together with equations (6.2) and (6.5), allows us to describe the behavior of this system with the second-order differential equation:

$$m\frac{d^2x}{dt^2} + a\frac{dx}{dt} + kx(t) = f(t) \tag{6.37}$$

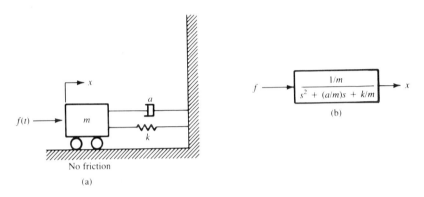

Figure 6.6   General second-order mechanical system.

or, in operator notation,

$$(mD^2 + aD + k)x(t) = f(t) \tag{6.38}$$

The roots of the corresponding characteristic equation are

$$r_1, r_2 = -\frac{a}{2m} \pm \left(\frac{a^2}{4m^2} - \frac{k}{m}\right)^{1/2} \tag{6.39}$$

The block diagram for this system is shown in Figure 6.6b, with the transfer function for the system given by

$$G(s) = \frac{x(s)}{f(s)} \tag{6.40}$$

$$= \frac{1}{ms^2 + as + k}$$

or

$$G(s) = \frac{1/m}{s^2 + (a/m)s + (k/m)} \tag{6.41}$$

The behavior of a general second-order system is described by the differential equation

$$\frac{d^2c}{dt^2} + 2\zeta\omega_n\frac{dc}{dt} + \omega_n^2 c(t) = \omega_n^2 r(t) \tag{6.42}$$

where

$c(t) \equiv$ general output or response function

$r(t) \equiv$ general input or forcing function

$\zeta \equiv$ damping ratio

$\omega_n \equiv$ undamped natural frequency

$r(t)$ can be a step function, a ramp function, an impulse function, and so on. (The physical significance of ζ and ω_n will be identified shortly.) The roots of the corresponding general characteristic equation are

$$r_1, r_2 = -\zeta\omega_n \pm j[\omega_n(1 - \zeta^2)^{1/2}] \tag{6.43}$$

where ζ has been assumed to be less than unity. The response function is then

$$c(t) = c_1(t) + 2e^{-\zeta\omega_n t}\{a \cos[\omega_n t(1 - \zeta^2)^{1/2}] + b \sin[\omega_n t(1 - \zeta^2)^{1/2}]\} \tag{6.44}$$

where $c_1(t)$ is that portion of $c(t)$ which is due to $r(t)$ [i.e., $c_1(t)$ is the steady-state solution of equation (6.42)].

For the case in which *no damping* occurs (i.e., no damping element—such as a dashpot, friction, or an electrical resistor—is contained within the system),

$$\zeta = 0 \tag{6.45}$$

so that

$$c(t) = c_1(t) + 2(a \cos \omega_n t + b \sin \omega_n t) \tag{6.46}$$

Therefore, the system's output oscillates about $c_1(t)$ with an (undamped natural) frequency ω_n.

If damping does occur,

$$\zeta \neq 0 \tag{6.47}$$

so that

$$c(t) \rightarrow c_1(t) + 0 \tag{6.48}$$

as $t \rightarrow \infty$; that is, the response function exponentially approaches the steady-state value $c_1(t)$ as t increases. Such an approach to $c_1(t)$ will include oscillatory behavior if $(0 < \zeta < 1)$.

If damping occurs such that

$$\zeta > 1 \tag{6.49}$$

then the roots of the general characteristic equation are

$$r_1, r_2 = -\zeta\omega_n \pm \omega_n(\zeta^2 - 1)^{1/2} \qquad (6.50)$$

which are purely real; as a result,

$$c(t) = c_1(t) + ae^{r_1 t} + be^{r_2 t} \qquad (6.51)$$

where r_1 and r_2 are negative real numbers. Therefore, $c(t)$ approaches $c_1(t)$ exponentially *without* oscillation.

In summary, the cases considered above (in terms of the value of the damping ratio ζ) are:

Case 1.   No damping; $c(t)$ oscillates about $c_1(t)$.

$$\zeta = 0 \qquad (3.45)$$

$$c(t) = c_1(t) + 2(a \cos \omega_n t + b \sin \omega_n t) \qquad (3.46)$$

Case 2.   Underdamping; $c(t)$ exponentially approaches $c_1(t)$ with oscillation.

$$0 < \zeta < 1 \qquad (6.52)$$

$$c(t) = c_1(t) + 2e^{-\zeta\omega_n t}\{a \cos [\omega_n t(1 - \zeta^2)^{1/2}] + b \sin [\omega_n t(1 - \zeta^2)^{1/2}]\} \qquad (6.53)$$

Case 3.   Critical damping; $c(t)$ exponentially approaches $c_1(t)$ without oscillation.

$$\zeta = 1 \qquad (6.54)$$

$$c(t) = c_1(t) + 2ae^{-\omega_n t} \qquad (6.55)$$

Case 4.   Overdamping; $c(t)$ exponentially approaches $c_1(t)$ *slowly* without oscillation.

$$\zeta > 1 \qquad (6.49)$$

$$c(t) = c_1(t) + ae^{r_1 t} + be^{r_2 t} \qquad (6.51)$$

with r_1 and r_2 given by equation (6.49).

Figure 6.7a shows $c(t)$ for each of these cases. Furthermore, the location of the roots r_1 and r_2 in the Laplace s-domain (that is, the complex s-plane) is dependent on the value of ζ, in accordance with the following considerations (Figure 6.7b):

Case 1.   No damping:

$$\zeta = 0$$

$$r_1, r_2 = \pm j\omega_n$$

(Both roots lie on the imaginary axis.)

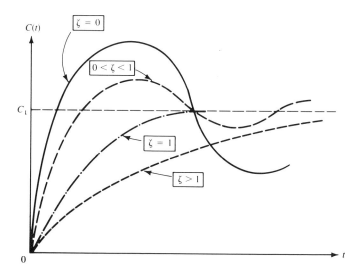

Figure 6.7a   Response behavior as a function of the damping ratio ζ.

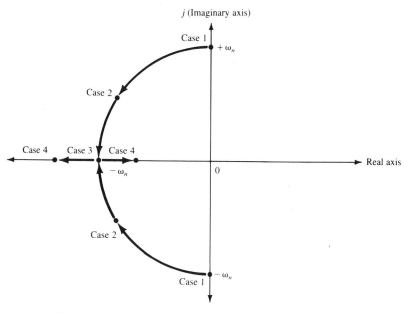

Case 1 No damping ($\zeta = 0$): $r_1 = j\omega_n$, $r_2 = -j\omega_n$
Case 2 Underdamping ($0 < \zeta < 1$): $r_1, r_2 = -\zeta\omega_n \pm j\omega_n (1 - \zeta^2)^{1/2}$
Case 3 Critical damping ($\zeta = 1$): $r_1, r_2 = -\omega_n$
Case 4 Overdamping ($\zeta > 1$): $r_1, r_2 = -\zeta\omega_n \pm (\zeta^2 - 1)^{1/2}$

Figure 6.7b   Location of the roots of the general second-order system in the complex s-plane in terms of the damping ratio ζ.

Case 2.   Underdamping:

$$0 < \zeta < 1$$

$$r_1, r_2 = -\zeta\omega_n \pm j\omega_n(1 - \zeta^2)^{1/2}$$

(Both roots lie in the complex s-plane; as ζ is increased, these roots approach the real axis.)

Case 3.   Critical damping:

$$\zeta = 1$$

$$r_1, r_2 = -\omega_n$$

(Both roots lie at the same location on the real axis.)

Case 4.   Overdamping:

$$\zeta > 1$$

$$r_1 = -\zeta\omega_n - \omega_n(\zeta^2 - 1)^{1/2}$$

$$r_2 = -\zeta\omega_n + \omega_n(\zeta^2 - 1)^{1/2}$$

(Both roots lie on the real axis.)

As the damping within the system is increased, the roots travel along a path (or a locus of root values) in the s-plane. (In chapter 8, we will consider root loci in stability analysis.) This path begins on the imaginary axis at root values $+j\omega_n$ and $-j\omega_n$ which correspond to the absence of damping ($\zeta = 0$) within the system. As damping is introduced, the real portions of the roots become nonzero (and negative); as a result, the transient portion of the system output is exponentially damped. For critical damping, both roots lie on the real axis (at $-\omega_n$); the transient portion of the system output is then exponentially damped *and* nonoscillatory in its behavior. If damping is further increased (overdamping), one root approaches $-\infty$ along the real axis, whereas the other root approaches zero along this axis (that is, as damping is increased above the critical point, one real root becomes increasingly negative while the other real root becomes less negative). The transient portion of the system output is damped most quickly at critical damping since *both* roots lie at the most negative real location $(-\omega_n)$ in the complex s-plane that they can achieve *simultaneously* for any value of the damping ratio ζ. For $0 < \zeta < 1$ *or* $\zeta > 1$, at least one root will include a real portion in its value that is less negative than $-\omega_n$; the term in the transient portion of the system output that corresponds to this root will then be more slowly exponentially damped than would be the case for $\zeta = 1$.

For the mechanical system of Figure 6.6, these cases correspond to the following situations:

$$\zeta = \frac{a}{2(km)^{1/2}} \tag{6.56}$$

$$\omega_n = \left(\frac{k}{m}\right)^{1/2} \tag{6.57}$$

[where we have simply equated coefficients of like terms in equations (6.37) and (6.42)]. The equation of motion for this system, in the Laplace domain, is

$$\left[s^2 + \left(\frac{a}{m}\right)s + \left(\frac{k}{m}\right)\right]x(s) = \frac{f(s)}{m} \tag{6.58}$$

The characteristic equation for this system has roots r_1 and r_2 given by

$$r_1, r_2 = -\frac{a}{2m} \pm \left(\frac{a^2}{4m^2} - \frac{k}{m}\right)^{1/2} \tag{6.59}$$

The cases to be considered are then:

1. *No damping:* In this case, there is no dashpot or frictional forces which dampen the effect of the applied force f (Figure 6.8a). Therefore,

$$a = 0 \tag{6.60}$$

The roots of the characteristic equation are then

$$r_1, r_2 = \pm j\left(\frac{k}{m}\right)^{1/2} \tag{6.61}$$

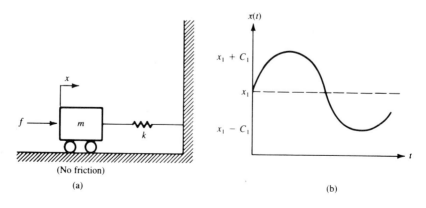

Figure 6.8 Mechanical system without damping.

[where $j = (-1)^{1/2}$], so that

$$x(t) = x_1(t) + C_1 \sin\left[\left(\frac{k}{m}\right)^{1/2} t + \phi\right]$$

(6.62)

that is, an undamped oscillation about the value $x_1(t)$ is predicted—as expected for a system in which energy is never dissipated (Figure 6.8b).

2. *Underdamped:* In this case (Figure 6.9a),

$$0 < \zeta = \frac{a}{2(km)^{1/2}} < 1$$

(6.63)

or

$$\frac{a^2}{m^2} < \frac{4k}{m}$$

(6.64)

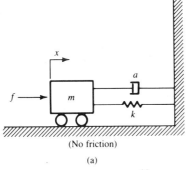

(No friction)

(a)

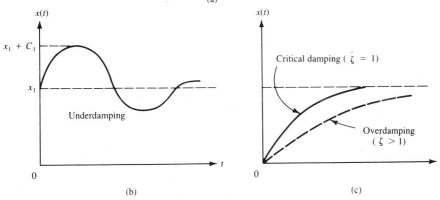

(b)

(c)

Figure 6.9   Mechanical system with damping.

Therefore, the roots r_1 and r_2 are *complex* numbers with

$$r_1 = A + jB$$

(6.65a)

$$r_2 = A - jB$$

(6.65b)

where

$$A = -\frac{a}{2m} \tag{6.66}$$

and

$$B = \frac{1}{2}\left(\frac{4k}{m} - \frac{a^2}{m^2}\right)^{1/2} \tag{6.67}$$

Thus

$$x(t) = x_1(t) + C_1 e^{At} \sin(Bt + \phi) \tag{6.68}$$

Since A is a negative number, $x(t)$ will oscillate about the value $x_1(t)$ with an amplitude C_1 which is exponentially damped with increasing time t (see Figure 6.9b).

3. *Critical damping:* In this case (Figure 6.9a),

$$\zeta = \frac{a}{2(km)^{1/2}}$$

$$= 1 \tag{6.69}$$

or

$$\frac{a^2}{m^2} = \frac{4k}{m} \tag{6.70}$$

so that

$$r_1, r_2 = -\frac{a}{2m} \pm \left(\frac{a^2}{m^2} - \frac{4k}{m}\right)^{1/2}$$

$$= -\frac{a}{2m} \pm 0 \tag{6.71}$$

that is, a double root exists at the value $-a/2m$. Therefore,

$$x(t) = x_1(t) + (C_1 + C_2 t)e^{-at/2m} \tag{6.72}$$

or

$$x(t) \rightarrow x_1(t)$$

as $t \rightarrow \infty$. The response $x(t)$ exponentially approaches the final value $x_1(t)$ without oscillation.

4. *Overdamped:* In this case, the damping effect of the dashpot is *more* than enough to eliminate oscillating behavior in $x(t)$, that is,

$$\zeta = \frac{a}{2(km)^{1/2}}$$

$$> 1 \tag{6.73a}$$

or

$$\frac{a^2}{m^2} > \frac{4k}{m} \tag{6.73b}$$

so that

$$r_1 = A + jB \tag{6.74a}$$

$$r_2 = A - jB \tag{6.74b}$$

where

$$A = -\frac{a}{2m} \tag{6.75a}$$

$$B = \left(\frac{a^2}{m^2} - \frac{4k}{m}\right)^{1/2} \tag{6.75b}$$

Both r_1 and r_2 are purely real numbers; thus

$$x(t) = x_1(t) + C_1 e^{r_1 t} + C_2 e^{r_2 t} \tag{6.76}$$

The response function $x(t)$ exponentially approaches its final value $x_1(t)$ more slowly than in the case of critical damping (see Figure 6.9c).

Figure 6.10 illustrates several characteristics of a sinusoidal response curve with which we may be concerned during system design (see Dorf, 1974):

- *Overshoot* is a measurement of the degree to which the response initially exceeds the desired output value C_1; in Figure 6.10, the ratio A/C_1 is the overshoot of the response.

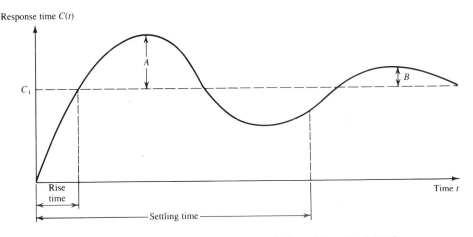

Figure 6.10 Sinusoidal response curve (Adapted from Dorf, 1974).

- *Decay ratio* indicates the amount of exponential decay between successive peaks of the response; in Figure 6.10, the decay ratio is A/B.
- *Rise time*, as its name implies, is the time interval between $t = 0$ and the initial achievement of the desired output value C_1.
- *Settling time* is the time interval required for the response to remain within $\pm \delta$ of the desired value C_1 [δ is often expressed as a percentage (5%, 2%) of C_1]; recall that after a time interval of 3τ, the system— with exponential damping—is within 5% of its final value and after an interval of 4τ, it is within 2% of its final value (if the system is stable and τ is the dominant time-delay constant for the system).

Example 6.3

Richardson proposed a mathematical model for an armament competition (or arms race) between two nations which can be represented by two first-order differential equations:

$$\frac{dx_1}{dt} = ax_2 - gx_1 + f_1$$

$$\frac{dx_2}{dt} = bx_1 - hx_2 + f_2$$

where

$x_1, x_2 \equiv$ armament levels (or the costs associated with such levels) for each nation

$a, b \equiv$ defense or reaction coefficients (reflecting the reaction of each nation to the size of the armament of its competitor)

$g, h \equiv$ fatigue or expense coefficients (denoting that economic constraints are imposed on the armament growth rate such that the rate decreases as the size of the armament increases)

$f_1, f_2 \equiv$ grievance coefficients (when positive); goodwill coefficients (when negative)

[see Richardson (1960) and Saaty and Alexander (1981)]. The quantities f_1 and f_2 act as driving forces to either increase or decrease the rate of armament growth, depending on international conditions.

If we apply Cramer's rule to these two first-order equations, we may develop a single second-order expression for x_1 or x_2; for example, consider

$$x_1 = \frac{\begin{vmatrix} f_1 & -a \\ f_2 & D+h \end{vmatrix}}{\begin{vmatrix} D+g & -a \\ -b & D+h \end{vmatrix}}$$

which can then be expressed in the form

$$[(D+g)(D+h) - ab]x_1 = (D+h)f_1 + af_2$$

or

$$[D^2 + (g+h)D + (gh - ab)]x_1 = (D+h)f_1 + af_2$$

(where $D = d/dt$). If we assume that both f_1 and f_2 are constants, a Laplace transformation then produces

$$[s^2 + (g + h)s + (gh - ab)]x_1(s) = \frac{hf_1 + af_2}{s}$$

(under the additional assumption of zero initial conditions). The characteristic equation for this system with respect to x_1 is then

$$s^2 + (g + h)s + (gh - ab) = 0$$

with roots r_1 and r_2 given by

$$r_1, r_2 = \frac{-(g + h) \pm [(g + h)^2 - 4(gh - ab)]^{1/2}}{2}$$

For purely real roots, the system parameters must obey the inequality

$$(g + h)^2 > 4(gh - ab)$$

Furthermore, if the system is to be stable (i.e., if the armament competition is to end with a stable balance of power existing between nations), we must have

$$-(g + h) + [(g + h)^2 - 4(gh - ab)]^{1/2} < 0$$

or, equivalently,

$$(g + h)^2 - 4(gh - ab) < (g + h)^2$$

which then leads to

$$ab - gh < 0$$

In other words, a final state of equilibrium will be achieved if

$$ab < gh$$

This inequality states that the armament competition will have an upper limit if the product of the *expense* coefficients is greater than the product of the *defense/reaction* coefficients. If the cost of armament construction exceeds the perceived need for national defense, a nation's armament will achieve a stable equilibrium value (at least, until events—domestic or international—alter the condition of the system).

Finally, we may obtain an expression for $x_1(t)$ by analyzing its Laplace transform $x_1(s)$:

$$x_1(s) = \frac{hf_1 + af_2}{s} \frac{1}{s^2 + (g + h)s + (gh - ab)}$$

$$= \frac{A_1}{s} + \frac{A_2}{s - r_1} + \frac{A_3}{s - r_2}$$

An inverse Laplace transformation then produces

$$x_1(t) = A_1 + A_2 e^{r_1 t} + A_3 e^{r_2 t}$$

For a stable system, r_1 and r_2 have negative real portions so that the two terms in $x_1(t)$ associated with these two roots vanish as $t \to \infty$. The coefficient A_1 can be shown to equal $(hf_1 + af_2)/r_1 r_2$. Thus

$$x_1(t \rightarrow \infty) = \frac{hf_1 + af_2}{r_1 r_2}$$

$$= \frac{hf_1 + af_2}{gh - ab}$$

We may also develop a relationship between system parameters for the case of critical damping. The characteristic equation for this system was shown to be

$$s^2 + (g + h)s + (gh - ab) = 0$$

If we now compare the coefficients of this expression with the general second-order characteristic equation given by

$$s^2 + 2\zeta\omega_n s + \omega_n^2 = 0$$

we obtain

$$\omega_n = (gh - ab)^{1/2}$$

and

$$\zeta = \frac{g + h}{2\omega_n}$$

$$= \frac{g + h}{2(gh - ab)^{1/2}}$$

Of course, for critical damping ζ must be equal to unity; the corresponding relationship between system parameters is then

$$2(gh - ab)^{1/2} - g + h$$

[The quantity $(g + h)$ corresponds to the total damping within the system due to economic constraints.]

6.5 SYSTEM COMPONENTS AND SYSTEM ANALOGIES

We have used various types of mechanical system components in our previous discussions. Let us now develop a brief review of the mechanical energy domain (see Ogata, 1978). Mechanical elements that can be used to construct a system include:

1. *Inertia (or capacitive storage) elements:* For translational motion, the mass M of a body represents its resistance to linear acceleration a, where

$$F = Ma = M\frac{d^2x}{dt^2} \tag{6.77}$$

For rotational motion, the moment of inertia J of a (rigid) body about an axis represents the resistance of the body to angular acceleration α about that axis. J is defined by the expression

$$J = \int r^2 \, dm \tag{6.78}$$

where r is the distance from the axis to the element of mass dm in the body and where the integration is performed over the entire mass of the rigid body.

2. *Spring (or inductive storage) elements:* Energy is stored in an inductive (spring) element as the element is deformed by an external applied force or torque. The amount of translational deformation x of linear springs is proportional to the applied force F, that is,

$$F = kx \tag{6.79}$$

where k is the spring constant of proportionality. In the case of rotational motion, the amount of angular deformation θ is proportional to the applied torque T:

$$T = \kappa\theta \tag{6.80}$$

where κ represents the torsional spring constant.

The compliance or mechanical inductance I_m is the reciprocal of the spring constant k:

$$I_m = \frac{1}{k} \tag{6.81}$$

It will be convenient to differentiate equation (6.79) with respect to time t, so that

$$\frac{dF}{dt} = k\frac{dx}{dt}$$

or

$$\begin{aligned} \frac{dx}{dt} &= \frac{1}{k}\frac{dF}{dt} \\ &= I_m\frac{dF}{dt} \end{aligned} \tag{6.82}$$

3. *Damper (energy dissipative) elements*: Damper elements are components that dissipate energy in the form of heat. For example, a translational damper or dashpot obeys the relation

$$F = a\frac{dx}{dt} = a\dot{x} \tag{6.83}$$

where F is the applied force, a the viscous damping coefficient and $\dot{x}$ $(= dx/dt)$ the rate of change in the position x of the dashpot. [Of course, we introduced such an element with equation (6.5); we include it here to provide a complete list of the basic mechanical components in one section.] Such a dashpot is essentially an oil-filled cylinder against which a piston works:

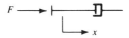

Heat is lost to the surroundings from the oil.

In the rotational case, a torsional damper obeys the relation

$$T = \beta \frac{d\theta}{dt} = \beta \dot{\theta} \tag{6.84}$$

where T is the applied torque, β the torsional viscous damping coefficient, and $\dot{\theta}$ the rate of change in the angular position θ of the dashpot.

In all of the components above, *ideal behavior* has been described; a real spring, for example, exhibits inertia (due to its finite mass) and damping (due to internal friction). However, we will assume that the behavior of each type of element can be approximated by its ideal behavior.

Furthermore, note that the viscous frictional behavior of mechanical dashpots, as described above, is limited to low speeds $\dot{x}$ and $\dot{\theta}$. The motion of a solid body in a fluid at *high* speeds produces frictional forces which are proportional to $\dot{x}^2$ or $\dot{\theta}^2$, in accordance with the square-law behavior shown in Figure 6.11 (see Ogata, 1978).

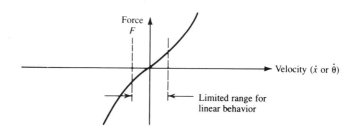

Figure 6.11   Square-law behavior (Adapted from Ogata, 1978).

Dry friction, in which unlubricated surfaces move relative to one another while in contact, can be subdivided into two categories:

1. *Static friction*: frictional force opposing the direction of motion *before* the object moves (i.e., while the body is static)

2. *Kinetic friction* (also known as Coulomb or sliding friction): frictional force applied to a body while in motion

Figure 6.12a illustrates dry frictional force as a function of velocity; note that the static friction achieves its maximum value just before the object begins moving. The static frictional force F_s is given by

$$F_s = \mu_s N \tag{6.85}$$

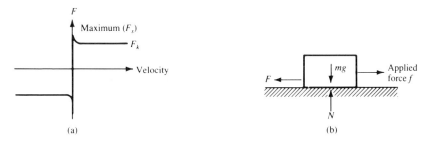

Figure 6.12   Frictional force acting upon a body (Adapted from Ogata, 1978).

where μ_s is the coefficient of static friction and N is the force normal to the surface along which the frictional force is applied (Figure 6.12b). Kinetic friction F_k is given by the expression

$$F_k = \mu_k N \tag{6.86}$$

where μ_k is the coefficient of kinetic friction. Furthermore, $\mu_s \geq \mu_k$ (see Ogata, 1978).

Similar types of elements can be defined in the *electrical* energy domain:

Capacitive elements, for which

$$i = C\frac{dV}{dt} \qquad V_2 \, \bullet \!\!\xrightarrow{\; i \;}\!\! | \, c \, | \!\!\longrightarrow\!\! \bullet \, V_1 \tag{6.87}$$

(V is the voltage across the element and i denotes current), where C is the capacitance of the element. Such an element may consist of two metal plates with a dielectric material between them.

Inductive elements, for which self-inductance is described by

$$V = L\frac{di}{dt} \qquad V_2 \, \bullet \!\!\xrightarrow{\; i \;}\!\! \text{mmm} \!\!\longrightarrow\!\! \bullet \, V_1 \tag{6.88}$$

where L represents the inductance of the element.

Resistive elements, for which

$$i = \frac{1}{R}V \cdot$$ $V_2 \bullet \!\!\!\!\xrightarrow{\quad i \quad}\!\!\!\!\!\text{\Large\char`\~}\!\bullet V_1$$ (6.89)

where R is the resistance of the element.

Compare the electrical relations above with the corresponding mechanical equations:

Capacitive storage:

$$F = M\frac{dx}{dt} \quad \text{(mechanical—translational)}$$

$$i = C\frac{dV}{dt} \quad \text{(electrical-capacitor)}$$

Inductive storage:

$$\dot{x} = C_m\frac{dF}{dt} \quad \text{(mechanical—spring)}$$

$$V = L\frac{di}{dt} \quad \text{(electrical—inductor)}$$

Resistive dissipation:

$$F = b\dot{x} \quad \text{(mechanical—dashpot)}$$

$$i = \frac{1}{R}V \quad \text{(electrical—resistor)}$$

Notice that if we define a generalized variable T (known as a *through* variable) and a generalized variable A (known as an *across* variable)—where we are following the presentation by Dorf (1974)—such that

T = through variable (electrical current i, mechanical force F, etc.)

A = across variable (electrical voltage V, translational velocity $\dot{x}$, etc.)

then we may write the generalized relations as follows:

Capacitive storage:

$$T = C\frac{dA}{dt}$$ (6.90)

[equivalent to equations (6.77) and (6.87)]. C is the generalized capacitance.

Inductive storage:

$$A = I\frac{dT}{dt}$$

(6.91)

[equivalent to equations (6.82) and (6.88)]. I is the generalized inductance.

Resistive dissipation:

$$T = \frac{1}{R}A$$

(6.92)

[equivalent to equations (6.83) and (6.89)]. R is the generalized resistance.

Table 6.1 gives these and other analogous relations for capacitive storage, inductive storage, and resistive dissipation elements in several energy domains. These analogous relations allow us to model a mechanical system, for example, with its equivalent electrical analog (which may be much more convenient to investigate experimentally than the actual mechanical system). Once the generalized equations in A and T for a system are solved, the solution applies to all systems (electrical, mechanical, thermal, etc.) which are analogs of the generalized model. Hence we may extend our results from a given type of system (e.g., mechanical) to all analogous systems. The importance of this generalization of results through analogs is significant!

Example 6.4

Consider the mechanical system shown in Figure 6.13a; the corresponding electrical system is given in Figure 6.13b.

One final point: the relations given in Table 6.1 form the basis of the *force-current analogy* (a particular set of equivalent relationships between system parameters in various energy domains); another set of equivalencies form the basis of the *force-voltage analogy* [see, e.g., Ogata (1978) and Dorf (1974)].

TABLE 6.1   Generalized Equations and the Corresponding
Relationships in Certain Energy Domains[a]

| | Energy Domain[b] | | | | |
Element	Mechanical–Translational	Mechanical–Rotational	Electrical	Fluid–Hydraulic	Generalized
Capacitive storage[c]	$F = M\dfrac{d\dot{x}}{dt}$	$\tau = J\dfrac{d\omega}{dt}$	$i = C_e\dfrac{dV}{dt}$	$Q = C_f\dfrac{dP}{dt}$	$T = C\dfrac{dA}{dt}$
Inductive storage[d]	$\dot{x} = I_m\dfrac{dF}{dt}$	$\omega = \kappa\dfrac{d\tau}{dt}$	$V = L\dfrac{di}{dt}$	$P = I_f\dfrac{dQ}{dt}$	$A = I\dfrac{dT}{dt}$
Dissipative[e]	$F = a\dot{x}$	$\tau = a\omega$	$i = \dfrac{1}{R_e}V$	$Q = \dfrac{1}{R_f}P$	$T = \dfrac{1}{R}A$

[a]These relationships form the basis for the force-current analogy (see also Ogata, 1978).

[b]*Through variables:* force F, current i, torque τ, fluid volumetric flow rate Q. *Across variables:* velocity $\dot{x}$, voltage V, angular velocity ω, pressure P.

[c]*Capacitive parameters:* mass M, electrical capacitance C_e, moment of inertia J, fluid capacitance C_f.

[d]*Inductive parameters:* reciprocal translational stiffness $I_m = k^{-1}$, inductance L, rotational stiffness κ, fluid inertance I_f.

[e]*Dissipative parameters:* viscous damping coefficient a, electrical resistance R, fluid resistance R_f.

Source: Adapted from Dorf (1974).

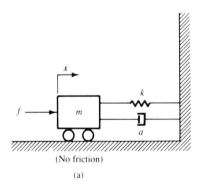

(No friction)

(a)

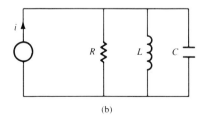

(b)

Figure 6.13 Analogous mechanical and electrical systems (Adapted from Ogata, 1978).

Example 6.5

Consider the mechanical system shown in Figure 6.14. An applied constant force acts on the system with initial conditions (at time $t = 0$) given by

$$x(t) = x(0) = 0$$
$$\dot{x}(t) = \dot{x}(0) = 0$$

(a) Determine the mathematical model that describes the behavior of this system.

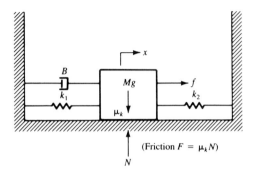

Values of system parameters for use in part (c) of Example 6.5:

$M = 5.0$	$f = A = 100.0$
$B = 2.0$	$\mu_k = 1.0$
$k_1 = 4.0$	$g = 9.81$ m/s^2
$k_2 = 3.0$	

Figure 6.14   System analyzed in Example 6.5.

Solution   The equations that relate the forces acting on the system are

$$N = Mg \qquad \text{(where } g \text{ is the acceleration due to gravity)}$$

$$\text{friction } F = \mu_k N = \mu_k Mg$$

$$M\frac{d^2x}{dt^2} = \Sigma f_i = \text{sum of all forces acting on the system}$$

so that

$$M\frac{d^2x}{dt^2} = f - \mu_k Mg - (k_1 + k_2)x - B\frac{dx}{dt}$$

or, upon transforming into the s-domain,

$$[Ms^2 + Bs + (k_1 + k_2)]x = (A - \mu_k Mg)\frac{1}{s} \tag{6.93}$$

where the applied constant force f is equal to A.

(b) Determine the value of the damping (dashpot) coefficient B for which the system is critically damped (in terms of the mass M and the spring constants k_1 and k_2).

Solution   The roots of the characteristic equation

$$Ms^2 + Bs + (k_1 + k_2) = 0$$

are given by

$$r_1, r_2 = -\frac{B}{2M} \pm \frac{1}{2M}[B^2 - 4(k_1 + k_2)M]^{1/2}$$

Critically damped behavior is achieved if the radical term is equal to zero, so that

$$B^2 = 4(k_1 + k_2)M$$

or

$$B = 2[(k_1 + k_2)M]^{1/2} \tag{6.94}$$

(c) Determine the response $x(t)$ of the system.

Solution   The Laplace transform $x(s)$ of the system response is obtained from equation (6.93):

$$x(s) = \frac{c}{s(Ms^2 + Bs + k_1 + k_2)}$$

$$= \frac{c/M}{s\left(s^2 + \dfrac{B}{M}s + \dfrac{k_1 + k_2}{M}\right)}$$

$$= \frac{E_1}{s} + \frac{E_2}{s - r_1} + \frac{E_3}{s - r_2}$$

where (for the values given in Figure 6.14)

$$c = A - \mu_k Mg = 50.95$$

Also, upon substitution of the given values for B, M, k_1, and k_2 into the expression for roots r_1 and r_2, we obtain

$$r_1, r_2 = -0.2 \pm j(1.36)^{1/2}$$

As a result, inverse transformation of $x(s)$ produces

$$x(t) = E_1 + Ce^{-0.2t}\sin(\omega t + \phi) \qquad [\text{where } \omega \equiv (1.36)^{1/2}]$$

$$E_1 = \frac{C}{k_1 + k_2} = 7.27857$$

At time t equal to zero:

$$x(0) = 0$$

$$= E_1 + C\sin\phi$$

$$\dot{x}(0) = 0$$

$$= -0.2Ce^{-0.2t}\sin(\omega t + \phi)$$

$$+ \omega Ce^{-0.2t}\cos(\omega t + \phi)$$

or

$$0 = -0.2C \sin \phi + \omega C \cos \phi$$

Therefore,

$$\tan \phi = \frac{\sin \phi}{\cos \phi} = \frac{\omega}{0.2} = \frac{(1.36)^{1/2}}{0.2}$$

or

$$\phi = \tan^{-1} \frac{(1.36)^{1/2}}{0.2} = 80.269$$

Also,

$$C \sin \phi = -E_1$$

or

$$C = \frac{-E_1}{\sin \phi} = \frac{-7.27857}{0.98561} = -7.3848$$

Finally, we have

$$x(t) = 7.27857 - 7.3848e^{0.2t} \sin(\omega t + 80.269) \tag{6.95}$$

Example 6.6

Figure 6.15 presents a simple electrical circuit consisting of a voltage source (of voltage v), a resistance R, and an inductance L (where i represents the current). The relationships given in Table 6.1 for elements from the electrical energy domain can then be

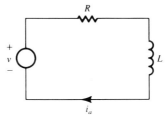

Figure 6.15   RL circuit.

used to represent the voltage drops across the resistance and inductance components, that is,

$$v_1 = iR$$

$$v_2 = L\frac{di}{dt}$$

Kirchhoff's voltage law states that the sum of the voltage changes (increases and decreases) around a closed loop is equal to zero; the RL circuit of Figure 6.15 can then be described by the following expression:

$$\sum_i v_i = v - v_1 - v_2$$

$$= v - iR - L\frac{di}{dt}$$

$$= 0$$

Rearrangement of this expression then provides us with the following first-order differential equation for the system:

$$L\frac{di}{dt} + iR = v$$

A Laplace transformation of this equation then produces

$$(Ls + R)i(s) = v(s)$$

where $i(s)$ and $v(s)$ are the Laplace transformations of $i(t)$ and $v(t)$, respectively (where we have assumed zero initial conditions). We may then generate an expression for $i(s)$; if we assume that the applied voltage $v(t)$ is constant (with a value equal to A), we obtain

$$i(s) = \frac{v(s)}{Ls + R}$$

$$= \frac{A}{s(Ls + R)}$$

$$= \frac{A_1}{s} + \frac{A_2}{Ls + R} = \frac{A_1}{s} + \frac{A_2/L}{s + (R/L)}$$

where we have expanded $i(s)$ in terms of partial fractions with unknown coefficients A_1 and A_2. An inverse transformation then produces an expression for $i(t)$:

$$i(t) = A_1 + \frac{A_2}{L}e^{-Rt/L}$$

We may also evaluate the coefficients A_1 and A_2, resulting in the expressions

$$A_1 = \frac{A}{R}$$

$$A_2 = \frac{-AL}{R}$$

so that

$$i(t) = \frac{A}{R}(1 - e^{-Rt/L})$$

Example 6.7

The *RLC* circuit shown in Figure 6.16 can be analyzed by using Kirchhoff's voltage law and the relationships given in Table 6.1. Each of the voltage drops around the

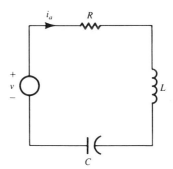

Figure 6.16   RLC circuit.

circuit are described by the expressions

$$v_1 = iR$$

$$v_2 = L\frac{di}{dt}$$

$$v_3 = \frac{1}{C}\int i(t)\, dt$$

The current i is related to the charge q flowing through the circuit in accordance with the expression

$$i \equiv \frac{dq}{dt}$$

Finally, Kirchhoff's voltage law states that the sum of voltage changes around the closed loop of the circuit must be equal to zero; therefore,

$$\sum_i v_i = v - v_1 - v_2 - v_3$$

$$= v - iR - L\frac{di}{dt} - \frac{1}{C}\int i(t)\, dt$$

$$= v - R\frac{dq}{dt} - L\frac{d^2q}{dt^2} - \frac{1}{C}(q)$$

$$= 0$$

A second-order expression in q may then be written:

$$L\frac{d^2q}{dt^2} + R\frac{dq}{dt} + \frac{1}{C}q = v$$

A Laplace transformation then gives

$$\left(Ls^2 + Rs + \frac{1}{C}\right)q(s) = v(s)$$

where we have assumed zero initial conditions. For simplicity, we will assume that the applied voltage v is equal to a constant A, so that we then obtain

$$q(s) = \frac{v(s)}{Ls^2 + Rs + C^{-1}}$$

$$= \frac{A}{s(Ls^2 + Rs + C^{-1})}$$

$$= \frac{A_1}{s} + \frac{A_2}{s - r_1} + \frac{A_3}{s - r_2}$$

where r_1 and r_2 represent the roots of the characteristic equation

$$Ls^2 + Rs + C^{-1} = 0$$

and where A_1, A_2, and A_3 are coefficients that need to be evaluated. An inverse transformation then produces an expression for $q(t)$:

$$q(t) = A_1 + A_2 e^{r_1 t} + A_3 e^{r_2 t}$$

where

$$A_1 = \frac{A}{r_1 r_2}$$

$$A_2 = \frac{A}{r_1(r_1 - r_2)}$$

$$A_3 = \frac{A}{r_2(r_2 - r_1)}$$

Example 6.8

An armature-controlled dc (direct-current) motor, driving an inertial load with damping, is an example of an important control engineering system which consists of components from both the *electrical* and *mechanical-rotational* energy domains (see Figure 6.17).

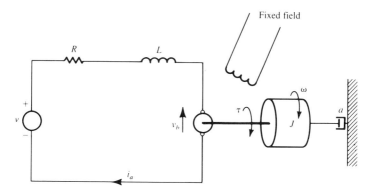

Figure 6.17 Armature-controlled dc motor with inertial load and damping (Adapted from Palm, 1983, Dorf, 1974, and Harrison and Bollinger, 1963).

Such dc motors are used in industrial robots (Luh, 1983). We wish to develop a model of this system which relates the armature voltage v (input)—which is used to control the motor—to the rotational velocity ω (output).

The armature of such a motor usually consists of an iron core about which is wound a wire conductor. An applied fixed magnetic field (supplied by an electromagnet with a constant voltage or by a permanent magnet) interacts with the armature current i_a to generate a torque which produces rotation of the armature. Rotation is then transferred to the inertial load for appropriate applications. The relevant variables and system parameters for this system include:

$$v \equiv \text{armature voltage}$$
$$i_a \equiv \text{constant armature current}$$
$$\omega \equiv \text{rotational velocity}$$
$$R \equiv \text{armature resistance}$$
$$L \equiv \text{inductance of armature winding}$$
$$a \equiv \text{viscous damping coefficient (rotational)}$$
$$J \equiv \text{moment of inertia}$$
$$v_b \equiv \text{back electromotive force (emf)}$$
$$\tau \equiv \text{torque}$$
$$K_T \equiv \text{torque constant}$$
$$K_e \equiv \text{voltage constant}$$

We may apply Kirchhoff's voltage law to the motor circuit of Figure 6.15; the applied voltage v must be equal to the sum of the voltage drops through the circuit, that is,

$$v = i_a R + L \frac{di_a}{dt} + v_b$$

where v_b represents the back electromotive force (emf) which opposes the voltage v. This induced voltage is proportional to the rotational velocity ω such that

$$v_b = K_e \omega$$

or

$$v = i_a R + L \frac{di_a}{dt} + K_e \omega$$

(In other words, we now have a relation in which both v and ω appear; however, we need additional equations to relate these two quantities properly.) The armature current i_a is proportional to the torque τ, or

$$\tau = K_T i_a$$

Finally, this torque is applied to both the inertial load and damping in accordance with the expression

$$\tau = J \frac{d\omega}{dt} + a\omega$$

(The reader may wish to compare the foregoing relations to the expressions given in Table 6.1 for individual system components.) Laplace transformations then produce the relations

$$v(s) = (R + Ls)i_a(s) + K_e\omega(s)$$

$$\tau(s) = K_T i_a(s)$$

$$= (Js + a)\omega(s)$$

(under the assumption of zero initial conditions), where $i_a(s)$, $v(s)$, $\tau(s)$, and $\omega(s)$ represent the Laplace transforms of these quantities. A final relationship between $v(s)$ and $\omega(s)$ can then be obtained from these equations:

$$v(s) = \frac{1}{K_T}[LJs^2 + (RJ + aL)s + (aR + K_e K_T)]\omega(s)$$

The entire system may then be represented by the equivalent open-loop block diagram:

$$v(s) \longrightarrow \boxed{K_T[LJs^2 + (RJ + aL)s + (aR + K_e K_T)]^{-1}} \longrightarrow \omega(s)$$

We now wish to develop the closed-loop feedback block diagram for this system. To accomplish this goal, we will focus on each system component individually, beginning with the armature. If we ignore the induced voltage v_b for the moment, we may represent the armature by a block diagram that relates the voltage v to the current i_a:

$$v(s) \longrightarrow \boxed{\frac{1}{R + Ls}} \longrightarrow i_a(s)$$

[which corresponds to the time-domain relation

$$v(t) = Ri_a + L\frac{di_a}{dt}]$$

The direct relation between the current i_a and the resultant torque τ can be represented by the following block:

$$i_a(s) \longrightarrow \boxed{K_T} \longrightarrow \tau(s)$$

These two block diagrams may then be combined to represent the armature within the system (without the induced voltage v_b):

$$v(s) \longrightarrow \boxed{\frac{1}{R + Ls}} \longrightarrow i_a(s) \longrightarrow \boxed{K_T} \longrightarrow \tau(s)$$

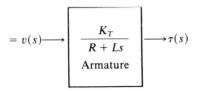

The inertial load and damping effect within the system can be represented by the block diagram

$$\tau(s) \rightarrow \boxed{\dfrac{1}{Js + a}} \rightarrow \omega(s)$$

Load

[which corresponds to the time-domain relation

$$\tau(t) = J\dfrac{d\omega}{dt} + a\omega(t)]$$

Finally, the (feed) back emf v_b can be related to the rotational speed ω in accordance with the block diagram

$$v_b(s) \leftarrow \boxed{K_e} \leftarrow \omega(s)$$

The entire closed-loop block diagram description of the system can then be constructed from these individual component diagrams (Figure 6.18). [The reader should demonstrate that the equivalent open-loop transfer function $T(s)$, which was obtained earlier, is

$$T(s) = \frac{K_T}{LJs^2 + (RJ + aL)s + (aR + K_e K_T)}$$

for this system.]

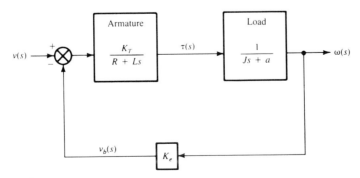

Figure 6.18   Block diagram for dc motor.

Our development of this example is based, in part, on Palm (1983), Dorf (1974), and Harrison and Bollinger (1963).

The characteristic equation for this system is then

$$LJs^2 + (RJ + aL)s + (aR + K_eK_T) = 0$$

or, upon division by LJ, so that the coefficient of the s^2 term is unity:

$$s^2 + \frac{RJ + aL}{LJ} + \frac{aR + K_eK_T}{LJ} = 0$$

One may then compare this expression to the general second-order characteristic equation,

$$s^2 + 2\zeta\omega_n s + \omega_n^2 = 0$$

in order to identify the damping ratio ζ and the undamped natural frequency ω in terms of the dc motor system parameters.

6.6 A MANAGEMENT-PRODUCTION SYSTEM

6.6.1 Model and Analysis

A management-production system can be modeled as a closed-loop feedback control system, as shown in Figure 6.19 (see Wilcox, 1963, and Dorf, 1974). The controller (i.e., management) is represented by a gain K_1, whereas the process (i.e., the production component) is mathematically described by a gain K_2 together with a time delay τ. The output $C(t)$ and its rate $\dot{C}(t)$ are measured in the feedback loop, which results in a corresponding response by the controller.

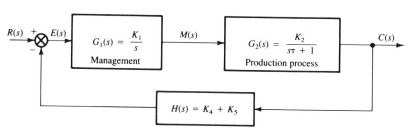

Figure 6.19   Management-production system (After Wilcox, 1963, and Dorf, 1974).

The general goals that one seeks in designing an appropriate control mechanism for such a system include:

- An accurate model of the system which correctly predicts the output response $C(t)$

- Maximum steady-state accuracy (or minimum steady-state error ϵ)
- Minimal settling time
- The achievement of the desired output response with minimal or no fluctuation or overshoot

(This list is not all-inclusive; in addition, all goals may not be achieved simultaneously or even sought in some situations. However, we will analyze this system in terms of our ability to achieve these goals.)

The equivalent open-loop transfer function $T(s)$ for this system is given by

$$T(s) = \frac{G_1 G_2}{1 + G_1 G_2 H}$$

$$= \frac{K_1 K_2 / \tau}{s^2 + [(1 + K_1 K_2 K_5)/\tau]s + (K_1 K_2 K_4 / \tau)}$$

$$= \frac{K}{s^2 + Bs + D}$$

where

$$B \equiv (1 + K_1 K_2 K_5)/\tau$$

$$D \equiv K_1 K_2 K_4 / \tau$$

$$K \equiv K_1 K_2 / \tau$$

We will restrict our attention to two possible types of input functions which represent demand acting on our system: a step input $R(t)$ equal to unity and a ramp input vt [where v is the slope of the ramp function and t represents the independent variable (time)]. (The reader may want to consider the physical interpretation of such demand inputs to the system.)

The steady-state error ϵ for this system is given by

$$\epsilon = 1 - \frac{1}{K_4}$$

for a unit step input and

$$\epsilon = \frac{v}{K_1 K_2 K_4}$$

for a ramp input equal to vt. If a step input is applied to the system, ϵ will vanish if K_4 is set equal to unity.

The roots of the characteristic equation will be *complex* or *purely real*, depending on the value of the control gain K_1. In particular, we find that:

1. *If the roots are complex,* the system is underdamped and an oscillatory response will result. We know that the system behavior will be unstable [i.e., the system response $C(t)$ will increase exponentially with time for any bounded input] if the real portion of the roots is positive; this can be seen from the general solution $C(t)$, which is (for the case of complex roots)

$$C(t) = \frac{RK}{b} \sin bt\, e^{at} \qquad \text{for a step input } R$$

or

$$C(t) = 2e^{at}(x \cos bt - y \sin bt) + L \qquad \text{for a ramp input } vt$$

where $a \equiv -B/2$

$b \equiv (4D - B^2)^{1/2}/2$

$L \equiv vK/(a^2 + b^2)$

with the two complex (conjugate) roots of the characteristic equation given by

$$r_1, r_2 = a \pm jb$$

2. *If the roots are purely real,* the system response is given by

$$C(t) = \frac{RK}{(B^2 - 4C)^{1/2}}(e^{r_1 t} - e^{r_2 t})$$

for a step input R or

$$C(t) = \frac{E}{r_1} e^{r_1 t} - \frac{E}{r_2} e^{r_2 t} + \frac{vK}{D}$$

[where

$$E \equiv \frac{vK}{(B^2 - 4C)^{1/2}} \Big]$$

for a ramp input vt. Again, it can be seen that the real roots r_1 and r_2 must be negative if the system is to be stable.

We now present a computer program with which the user (management or the controller) can determine the value of the gain K_1 needed to produce the results desired.

6.6.2 Software and Analysis

With the software described in Table 6.2a (in which a complete program listing is included), the user may evaluate the management-production system

TABLE 6.2a   Program Listing for MANPRO for Analysis of the Management-Production System Shown in Figure 6.15

```
C
C      ...A PROGRAM TO COMPUTE AND GRAPHICALLY DISPLAY THE OUTPUT OF A
C      MANAGEMENT-PRODUCTION SYSTEM MODELED BY A SECOND-ORDER
C      FEEDBACK CONTROL LOOP.
C
       IMPLICIT    REAL(K)
       CHARACTER*4 IN(2), DEMAND
       CHARACTER   ROOT*7, STABLE*1
       REAL        OUT(3,52), TIME(52)
C
       DATA IN,STABLE/'STEP','RAMP','Y'/
C
       OPEN(6,FILE='OUT.TXT')
C
       WRITE(*,100)
C
       WRITE(*,*) ' STEP?  RAMP?  K1MIN?   K1MAX? '
       READ(*,*)  STEP, RAMP, K1MIN, K1MAX
C
       WRITE(*,*) ' K3?   K4?    K5? '
       READ(*,*)  K3, K4, K5
C
       WRITE(*,*) ' TAU?  TMIN?  TMAX?   DELT? '
       READ(*,*)  TAU, TMIN, TMAX, DELT
C
       WRITE(6,105) K3, K4, K5, TAU
C
       DO 10 J = 1,2
          DEMAND = IN(J)
          K1     = K1MIN
          DELK1  = (K1MAX - K1MIN) / 2.
C
          NT  = IFIX( (TMAX - TMIN)/DELT + 1. )
C
          DO 20 LK1 = 1, 3
             K2  = .5 * K1 + K3
             A   = K1 * K2 / TAU
             B   = (1. + K1 * K2 * K5) / TAU
             C   = K1 * K2 * K4 / TAU
C
C   DETERMINE WHETHER ROOTS ARE REAL OR COMPLEX
C
             BSQ = B**2
             C4 = 4. *  C
C
             IF (BSQ .LT. C4) THEN
                ROOT = 'COMPLEX'
                X   = -B / 2.
                Y   = SQRT(C4 - BSQ) / 2.
             ELSE
                ROOT = 'REAL'
                X   = -B / 2.
                Y   = SQRT(BSQ - C4) / 2.
                R1  = X + Y
                R2  = X - Y
             END IF
C
C   CHECK STABILITY CRITERIA
C
```

TABLE 6.2a   Program Listing for MANPRO (continued)

```
      IF (ROOT .EQ. 'COMPLEX' .AND. X .GT. 0.) THEN
        STABLE = 'N'
        WRITE(6,*)   '  SYSTEM IS UNSTABLE!'
      ELSE IF (ROOT .EQ. 'REAL' .AND.
     1          (R1 .GT. 0. .OR. R2 .GT. 0.)) THEN
        STABLE = 'N'
        WRITE(6,*)   '  SYSTEM IS UNSTABLE!'
      END IF
C
      IF (DEMAND .EQ. 'STEP') THEN
        WRITE(6,110) K1, DEMAND, STEP, ROOT
      ELSE
        WRITE(6,110) K1, DEMAND, RAMP, ROOT
      END IF
C
      T = TMIN
C
      IF (STABLE .EQ. 'Y') THEN
C
      DO 30 LT = 1, NT
C
        TIME(LT) = T
C
        IF (ROOT .EQ. 'COMPLEX' .AND. DEMAND .EQ. 'STEP') THEN
        U        = (STEP * A)/C
        V        = -(STEP * A)/(X**2 + Y**2)
        W        = -V * (X / Y)
        OUT(LK1,LT) = U + EXP(X*t) * (V * COS(Y*T)
     1                   + W * SIN(Y*T))
      ELSE IF (ROOT .EQ. 'COMPLEX' .AND.
     1           DEMAND .EQ. 'RAMP') THEN
        U        = (RAMP * A)/C
        V        = -U * (B / C)
        W        = (U + V * (1.+X)) / Y
        OUT(LK1,LT) = U*T + V * (1. - EXP(X*T) * COS(Y*T))
     1                   - W * EXP(X*T) * SIN(Y*T)
      ELSE IF (ROOT .EQ. 'REAL' .AND. DEMAND .EQ. 'STEP') THEN
        U        = (STEP * A)/C
        V        = (STEP * A) / (R1 * (R1-R2))
        W        = (STEP * A) / (R2 * (R2-R1))
        OUT(LK1,LT) = U + V * EXP(R1*T) + W * EXP(R2*T)
      ELSE IF (ROOT .EQ. 'REAL' .AND. DEMAND .EQ. 'RAMP') THEN
        U        = (RAMP * A)/C
        V        = -U * (B / C)
        W        = (U + V * (1.+R1)) / (R2 - R1)
        OUT(LK1,LT) = U*T + V * (1. - EXP(R2*T))
     1                   + W * (EXP(R1*T) - EXP(R2*T))
      END IF
C
C     PRINT OUTPUT AND DYNAMIC ERROR ( = OUTPUT - INPUT) FOR ALL TIME
C
        IF (DEMAND .EQ. 'STEP') THEN
        DERROR = OUT(LK1,LT) - STEP
        WRITE(6,120)  T, STEP, OUT(LK1,LT), DERROR
      ELSE IF (DEMAND .EQ. 'RAMP') THEN
        RAMPT  = RAMP * T
        DERROR = OUT(LK1,LT) - RAMPT
        WRITE(6,120)  T, RAMPT, OUT(LK1,LT), DERROR
      END IF
C
```

TABLE 6.2a   Program Listing for MANPRO (continued)

```
        T = T + DELT
C
   30       CONTINUE
C
        END IF
C
C    SETTLING TIME   (ST)
C
        IF (ROOT .EQ. 'COMPLEX') THEN
           ST = 8. / B
        ELSE IF (BSQ .EQ. C4) THEN
           ST = 4. / SQRT(C)
           WRITE(6,*)
           WRITE(6,*)   ' SYSTEM HAS ACHIEVED CRITICAL DAMPING!'
        ELSE IF (R1 .LT. R2 .AND. R2 .LT. 0.) THEN
           ST = -4. / R2
        ELSE IF (R2 .LT. R1 .AND. R1 .LT. 0.) THEN
           ST = -4. / R1
        END IF
C
        WRITE(6,*)
        WRITE(6,*)   '  SETTLING TIME = ',ST
C
        STABLE = 'Y'
        K1     = K1 + DELK1
C
   20       CONTINUE
C
   10 CONTINUE
C
        CLOSE(6,STATUS='KEEP')
C
  100 FORMAT(//,T4,'ENTER INPUT DATA FOR THE PROGRAM:',/)
  105 FORMAT(//,T4,'MANPRO:    K3 = ',F4.2,',   K4 = ',F3.1,',   K5 = ',
     1       F3.1,',   TAU = ',F3.1)
  110 FORMAT(//,T4,'K1 = ',F4.2,' : ',A4,' INPUT = ',F3.1,', ',A7,
     1       ' ROOTS',//,T4,'TIME',4X,' INPUT',4X,' OUTPUT',4X,
     2       ' DY ERR',/)
  120 FORMAT(T4,F4.1,4X,F6.2,2(4X,F7.3))
C
        END
```

shown in Figure 6.15. An input data file must be created in which one provides a proposed step input, a proposed ramp input (the program evaluates both types of inputs), a minimum value for K_1, a maximum value for K_1, a proposed starting value for the time (t_{min}), a maximum value for time (t_{max}), an incremental change in time (Δt), and initial values for K_3, K_4, K_5, and τ.

The program will then determine if the system is stable in design; in addition, the K_1 values corresponding to both minimum and maximum settling times will be determined. A table of output results will also be generated in which—for each time t during the analysis—values for the step function (or ramp function) acting as input, the output or response of the system, and the dynamic error (defined as the difference between the desired and actual outputs of the system at any time during the analysis) are presented (see Tables 6.2b and 6.2c). (The dynamic error is identical to the steady-state error once the system has achieved equilibrium, that is, once the system is in a steady-state mode.)

Graphical descriptions of the system response for various K_1 values are presented in Figures 6.20 and 6.21.

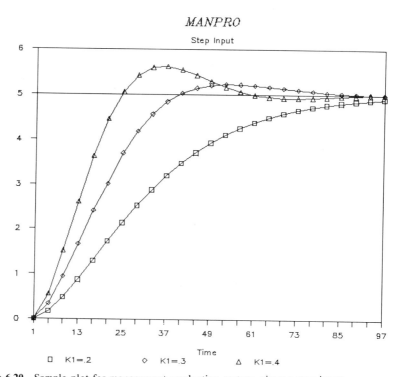

Figure 6.20 Sample plot for management-production system, given a step input.

TABLE 6.2b   MANPRO Program Test Results for a Given Step Input

---

MANPRO: K3 =   .05,   K4 = 1.0,   K5 =   .5,   TAU = 9.0

K1 =   .20 : STEP INPUT = 5.0, COMPLEX ROOTS

TIME	INPUT	OUTPUT	DY ERR
1.0	5.00	.008	-4.992
5.0	5.00	.173	-4.827
9.0	5.00	.484	-4.516
13.0	5.00	.876	-4.124
17.0	5.00	1.301	-3.699
21.0	5.00	1.730	-3.270
25.0	5.00	2.143	-2.857
29.0	5.00	2.528	-2.472
33.0	5.00	2.879	-2.121
37.0	5.00	3.193	-1.807
41.0	5.00	3.471	-1.529
45.0	5.00	3.713	-1.287
49.0	5.00	3.922	-1.078
53.0	5.00	4.101	-.899
57.0	5.00	4.254	-.746
61.0	5.00	4.383	-.617
65.0	5.00	4.491	-.509
69.0	5.00	4.582	-.418
73.0	5.00	4.658	-.342
77.0	5.00	4.721	-.279
81.0	5.00	4.773	-.227
85.0	5.00	4.816	-.184
89.0	5.00	4.851	-.149
93.0	5.00	4.880	-.120
97.0	5.00	4.903	-.097

SETTLING TIME =   70.9359600

K1 =   .30 : STEP INPUT = 5.0, COMPLEX ROOTS

TIME	INPUT	OUTPUT	DY ERR
1.0	5.00	.016	-4.984
5.0	5.00	.343	-4.657
9.0	5.00	.945	-4.055
13.0	5.00	1.669	-3.331
17.0	5.00	2.410	-2.590
21.0	5.00	3.099	-1.901
25.0	5.00	3.696	-1.304
29.0	5.00	4.184	-.816
33.0	5.00	4.562	-.438
37.0	5.00	4.839	-.161
41.0	5.00	5.027	.027
45.0	5.00	5.144	.144
49.0	5.00	5.206	.206
53.0	5.00	5.228	.228
57.0	5.00	5.222	.222
61.0	5.00	5.200	.200
65.0	5.00	5.168	.168
69.0	5.00	5.135	.135
73.0	5.00	5.102	.102
77.0	5.00	5.072	.072
81.0	5.00	5.047	.047

TABLE 6.2b   MANPRO Results for Step Input (continued)

85.0	5.00	5.028	.028
89.0	5.00	5.013	.013
93.0	5.00	5.002	.002
97.0	5.00	4.996	-.004

SETTLING TIME =   69.9029200

$K1 = .40$: STEP INPUT = 5.0, COMPLEX ROOTS

TIME	INPUT	OUTPUT	DY ERR
1.0	5.00	.027	-4.973
5.0	5.00	.564	-4.436
9.0	5.00	1.522	-3.478
13.0	5.00	2.610	-2.390
17.0	5.00	3.626	-1.374
21.0	5.00	4.456	-.544
25.0	5.00	5.053	.053
29.0	5.00	5.419	.419
33.0	5.00	5.591	.591
37.0	5.00	5.615	.615
41.0	5.00	5.545	.545
45.0	5.00	5.424	.424
49.0	5.00	5.289	.289
53.0	5.00	5.164	.164
57.0	5.00	5.063	.063
61.0	5.00	4.990	-.010
65.0	5.00	4.946	-.054
69.0	5.00	4.926	-.074
73.0	5.00	4.924	-.076
77.0	5.00	4.933	-.067
81.0	5.00	4.948	-.052
85.0	5.00	4.965	-.035
89.0	5.00	4.980	-.020
93.0	5.00	4.993	-.007
97.0	5.00	5.002	.002

SETTLING TIME =   68.5714300

TABLE 6.2c   MANPRO Program Test Results for a Given Ramp Input.

MANPRO: $K3 = .05$, $K4 = 1.0$, $K5 = .5$,   TAU = 9.0

$K1 = .30$: RAMP INPUT = 2.0, COMPLEX ROOTS

TIME	INPUT	OUTPUT	DY ERR
1.0	2.00	28.699	26.699
5.0	10.00	112.828	102.828
9.0	18.00	157.358	139.358
13.0	26.00	173.735	147.735
17.0	34.00	171.590	137.590
21.0	42.00	158.587	116.587
25.0	50.00	140.496	90.496
29.0	58.00	121.399	63.399
33.0	66.00	103.964	37.964
37.0	74.00	89.739	15.739
41.0	82.00	79.429	-2.571
45.0	90.00	73.150	-16.850
49.0	98.00	70.633	-27.367
53.0	106.00	71.388	-34.612

TABLE 6.2c   MANPRO Results for Ramp Input (continued)

57.0	114.00	74.828	-39.172
61.0	122.00	80.353	-41.647
65.0	130.00	87.408	-42.592
69.0	138.00	95.510	-42.490
73.0	146.00	104.267	-41.733
77.0	154.00	113.374	-40.626
81.0	162.00	122.608	-39.392
85.0	170.00	131.817	-38.183
89.0	178.00	140.909	-37.091
93.0	186.00	149.834	-36.166
97.0	194.00	158.575	-35.425

```
SETTLING TIME =            69.9029200

K1 =   .40 : RAMP INPUT = 2.0, COMPLEX ROOTS
```

TIME	INPUT	OUTPUT	DY ERR
1.0	2.00	17.480	15.480
5.0	10.00	67.479	57.479
9.0	18.00	90.831	72.831
13.0	26.00	95.324	69.324
17.0	34.00	88.482	54.482
21.0	42.00	76.608	34.608
25.0	50.00	64.362	14.362
29.0	58.00	54.716	-3.284
33.0	66.00	49.171	-16.829
37.0	74.00	48.093	-25.907
41.0	82.00	51.087	-30.913
45.0	90.00	57.335	-32.665
49.0	98.00	65.870	-32.130
53.0	106.00	75.763	-30.237
57.0	114.00	86.243	-27.757
61.0	122.00	96.744	-25.256
65.0	130.00	106.908	-23.092
69.0	138.00	116.558	-21.442
73.0	146.00	125.655	-20.345
77.0	154.00	134.251	-19.749
81.0	162.00	142.449	-19.551
85.0	170.00	150.371	-19.629
89.0	178.00	158.130	-19.870
93.0	186.00	165.821	-20.179
97.0	194.00	173.512	-20.488

```
SETTLING TIME =            68.5714300

K1 =   .50 : RAMP INPUT = 2.0, COMPLEX ROOTS
```

TIME	INPUT	OUTPUT	DY ERR
1.0	2.00	11.869	9.869
5.0	10.00	44.892	34.892
9.0	18.00	58.203	40.203
13.0	26.00	58.189	32.189
17.0	34.00	51.572	17.572
21.0	42.00	43.892	1.892
25.0	50.00	38.803	-11.197
29.0	58.00	38.036	-19.964
33.0	66.00	41.766	-24.234
37.0	74.00	49.171	-24.829
41.0	82.00	58.979	-23.021
45.0	90.00	69.901	-20.099

TABLE 6.2c   MANPRO Results for Ramp Input (continued)

49.0	98.00	80.909	-17.091
53.0	106.00	91.346	-14.654
57.0	114.00	100.924	-13.076
61.0	122.00	109.642	-12.358
65.0	130.00	117.677	-12.323
69.0	138.00	125.283	-12.717
73.0	146.00	132.703	··13.297
77.0	154.00	140.128	-13.872
81.0	162.00	147.676	-14.324
85.0	170.00	155.394	-14.606
89.0	178.00	163.275	-14.725
93.0	186.00	171.283	-14.717
97.0	194.00	179.367	-14.633

SETTLING TIME =   66.9767500

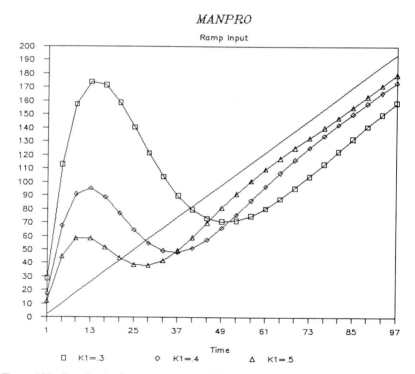

Figure 6.21   Sample plot for management-production system, given a ramp input.

6.7 CONTINUOUS SYSTEM SIMULATION

Systems in which smooth changes in variable values occur are known as *continuous*. Differential equations are used to model continuous systems and monitor the *rates of change* in system variables. As noted in this chapter, processes in a wide range of different energy domains may be represented by differential equations; in addition, growth rates in socioeconomic systems may be investigated through the use of differential equations (see examples 6.1 and 6.2). As Gordon (1978) has indicated, the analysis of continuous systems can be both extremely laborious and difficult (particularly if nonlinear models are used). To facilitate such analysis, continuous system simulation software may be used by the investigator.

Continuous system simulation software allows one to generate numerical or graphical representations of system output values at specified times during the simulated process. The system model may be altered to investigate the effects of organizational structure, specific parameter values, and various forcing functions on system behavior. Continuous system simulation languages include **CSMP** (an acronym for Continuous System Modeling Program; see Speckhart and Green, 1976); **DYNAMO** (developed for the simulation of socioeconomic systems; see Pugh, 1973); and **TUTSIM** (a simulation language for microcomputers in which both bond graphs and block diagrams may be used to model a system; see TUTSIM, 1984). Gordon (1978) provides an extended discussion of continuous system simulation. Additional references include Jacoby and Kowalik (1980) and Deo (1983).

One final word about system simulation: The analyst must be able to model the system under investigation properly before he or she can fully apply simulation software in the analysis of the system behavior. In other words, simulation software is dependent on the user's mastery of the material in this book.

6.8 REVIEW

In summary, we have reviewed the following topics, facts, relationships, or concepts in this chapter.

- Zeroth-order, first-order, general first-order, and general second-order systems were analyzed. Transfer functions for each type of system were developed.
- The response of zeroth-order, first-order, and so on, systems to various types of forcing functions (step, impulse, sinusoidal) was evaluated. In particular, the response of a system as a function of the *damping ratio ζ* was investigated, leading to the specific cases in which the

system is *undamped* ($\zeta = 0$), *underdamped* ($0 < \zeta < 1$), *critically damped* ($\zeta = 1$), or *overdamped* ($\zeta > 1$).

- The concepts of *decay ratio, overshoot, rise time,* and *settling time* were introduced, together with the significance of the time constant τ for the system under consideration.
- Energy domains (e.g., mechanical, electrical, thermal) were introduced, together with the concept of energy storage elements (either capacitive or inductive) and energy dissipation elements. The method of *system analogies* was then reviewed, in which equivalent systems in different energy domains can be represented mathematically by equations of identical form; one may then adapt a solution for one type of system (e.g., mechanical) as the response of an analogous system in a different energy domain by simply equating analogous variables and system parameters. Variables were categorized as either *through* or *across*; parameters are *dissipative* (resistive), *inductive*, or *capacitive* in nature. *Generalized equations* were also given.

EXERCISES

6.1. Consider the system shown below, with $M = 2$, $a = 3$ and $f(t) = 2e^{-t}$.

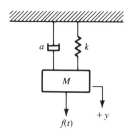

(a) Obtain the differential equation that describes this system.
(b) Obtain the Laplace transform of the differential equation that you found in part (a).
(c) Determine the value of the spring constant k for which critical damping is achieved.
(d) Determine the system output $y(t)$ for critical damping and zero initial conditions.

6.2. Consider the system shown in Exercise 6.1, with the parameter values $M = 10$, $k = 4$, and $f(t) = 30\,\delta t$ (i.e., an impulse function).
(a) Obtain the differential equation that describes this system.
(b) Obtain the Laplace transform of the differential equation that you found in part (a).
(c) Determine the value of the damping coefficient a for which critical damping is achieved.

(d) Determine the system output $y(t)$ for zero initial conditions and an unknown damping coefficient a.

(e) Plot $y(t)$ as a function of t for $a = 0.1$, $a = 1$, $a = 10$, and $a = 100$. Choose an appropriate range for the independent variable t.

6.3. Given the first-order system

with corresponding differential equation

$$(2D + 1)x(t) = 4$$

where $D = d/dt$ with zero initial conditions:

(a) Determine x_{ss}, the steady-state portion of $x(t)$.

(b) Determine the time constant τ for this system.

(c) Evaluate $x(t)$ at the specific times $t = \tau$, $t = 2\tau$, $t = 3\tau$, $t = 4\tau$, and $t = 5\tau$.

6.4. For the system shown in Exercise 6.1, use the system parameters $M = 10$, $k = 4$, and $f(t) = 15 \sin 2t$.

(a) Determine the natural frequency of the response $y(t)$.

(b) Do you expect the amplitude of the response to be greater than that of $f(t)$? Explain your reasoning.

(c) Determine the response $y(t)$ for the case where $a = 2$. Assume zero initial conditions.

6.5. Consider the system shown below (assume that frictional effects can be neglected).

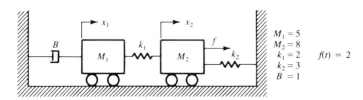

(a) Obtain the differential equations that describe the behavior of this system.

(b) Use Cramer's rule to obtain a single differential equation in which only x_1 appears as an unknown dependent variable.

(c) Obtain an equivalent open-loop block diagram for this system in which $x_1(s)$ is the output and $f(s)$ denotes the input. Include an expression for the transfer function $T(s)$.

(d) Determine the system output $x_1(t)$. Assume zero initial conditions.

6.6. Consider the system shown below (assume that frictional effects can be neglected).

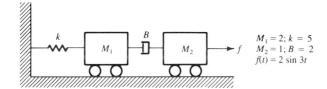

(a) Obtain the differential equations that describe the behavior of this system.

(b) Use Cramer's rule to obtain a single differential equation in which only x_1 appears as an unknown dependent variable.

(c) Evaluate the equation obtained in part (b) for $x_1(t)$. Assume zero initial conditions.

6.7. Consider the system shown in Exercise 6.6; however, include frictional effects with $\mu_k = 2.0$. Perform the analysis of this system with frictional effects in accordance with the directions given in parts (a), (b), and (c) of Exercise 6.6.

6.8. Develop an electrical system that is equivalent to the mechanical system described in Exercise 6.5.

6.9. Develop an electrical system that is equivalent to the mechanical system described in Exercise 6.6.

6.10. Consider the following electrical system:

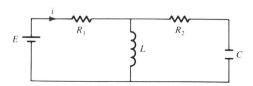

(a) Develop a mechanical system that is equivalent to the electrical system above.

(b) Write the generalized equations that describe the behavior of systems analogous to the electrical system above.

(c) Solve the foregoing generalized mathematical description of systems analogous to the given electrical system for the appropriate generalized variable. Write the corresponding solution to the electrical system given above.

6.11. Consider the following mechanical system:

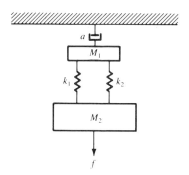

(a) Develop an electrical system that is equivalent to the mechanical system above.

(b) Write the generalized equations that describe the behavior of systems analogous to the mechanical system above.

(c) Solve the foregoing generalized mathematical description of systems analogous to the given mechanical system for the appropriate generalized variable. Write the corresponding solution to the mechanical system given above.

6.12. For the electrical system given in Exercise 6.10, develop the analogous mechanical system and obtain the transfer function $T(s)$ for the equivalent open-loop system in generalized notation.

6.13. Given the following differential equation (in operator notation) for a mechanical system:

$$(3D^2 + 56D + 4)x(t) = 3e^{-5t} + 4t$$

(a) Determine the damping ratio ζ and the undamped natural frequency ω_n for this system.

(b) Determine the generalized output $c(t)$ for the generalized system that is analogous to this mechanical system.

6.14. Consider the following mechanical system.

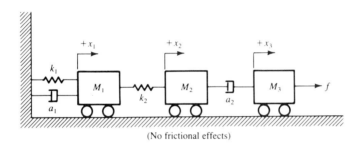

(No frictional effects)

(a) Develop the direct electrical analog for this system.

(b) Develop the direct rotational-mechanical analog for this system.

(c) Obtain the differential equations that describe the behavior of this mechanical system.

(d) Obtain the differential equations that describe—in generalized variable notation—the analog systems of this mechanical system.

6.15. A simple mechanical system is shown below.

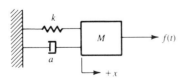

(a) Develop the differential equation that describes the behavior of this system.

(b) For the system parameters $M = 4$, $k = 3$, $a = 2$, and $f(t) = 3 \sin 5t$, together with zero initial conditions, evaluate this system and obtain $x(t)$.

6.16. A generalized system has the following parameters:

$$\zeta = 100$$

$$\omega = 60$$

$$r(t) = 4 \sin 10t$$

  (a) Obtain the general output or response function $c(t)$ for this system.
  (b) Obtain $c(t)$ for the case in which there is no damping ($\zeta = 0$).

6.17. Consider the following mechanical system:

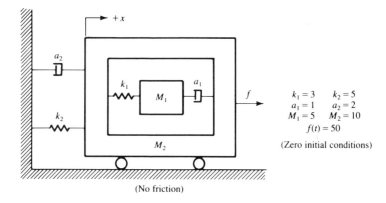

$$\begin{array}{ll} k_1 = 3 & k_2 = 5 \\ a_1 = 1 & a_2 = 2 \\ M_1 = 5 & M_2 = 10 \\ & f(t) = 50 \end{array}$$

(Zero initial conditions)

(No friction)

  (a) Develop the differential equation that describes the system's behavior.
  (b) Use Laplace transformations to obtain an expression for the system response $x(t)$.
  (c) Write a computer program that will determine the (1) decay ratio, (2) overshoot, (3) rise time, and (4) settling time (based on a maximum variation of 5% of the desired output response or steady-state behavior) for the system. The program should also produce a plot of $x(t)$ as a function of the time t.
  (d) Determine the value of a_1 that will produce critical damping for the given values of the other system parameters.

6.18. Consider the following mechanical system:

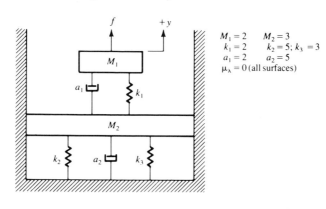

$$\begin{array}{ll} M_1 = 2 & M_2 = 3 \\ k_1 = 2 & k_2 = 5; k_3 = 3 \\ a_1 = 2 & a_2 = 5 \\ \mu_\lambda = 0 \, (\text{all surfaces}) \end{array}$$

(a) Develop the differential equation that describes the behavior of the mass M_1 [i.e., an equation in which only $y_1(t)$ appears as a dependent variable].

(b) Use Laplace transformations to obtain an expression for $y_1(t)$.

(c) Write a computer program that will determine the (1) decay ratio, (2) overshoot, (3) rise time, and (4) settling time (based on a maximum variation of 2% about the equilibrium response) for the system. The program should also produce a plot of $y_1(t)$ as a function of the time t.

(d) Determine the value of a_1 that will produce critical damping for the given values of the other system parameters.

6.19. Develop the electrical system that is the (force-current) analog of the mechanical system shown in Exercise 6.17.

6.20. Develop the electrical system that is the (force-current) analog of the mechanical system shown in Exercise 6.18.

6.21. Develop the fluid (hydraulic) system that is the analog of the mechanical system shown in Exercise 6.17.

6.22. Develop the fluid (hydraulic) system that is the analog of the mechanical system shown in Exercise 6.18.

6.23. Develop the rotational–mechanical system that is the analog of the translational–mechanical system shown in Exercise 6.17.

6.24. Develop the rotational–mechanical system that is the analog of the translational–mechanical system shown in Exercise 6.18.

6.25. Develop the electrical system that is the analog of the mechanical system shown in Exercise 2.19.

6.26. Develop the electrical system that is the analog of the mechanical system shown in Exercise 4.21.

6.27. Write the generalized equations (in terms of through and across variables) which describe the behavior of the generalized system that is analogous to the mechanical system shown in Exercise 6.17.

6.28. Write the generalized equations (in terms of through and across variables) which describe the behavior of the generalized system that is analogous to the mechanical system shown in Exercise 6.18.

6.29. Write the generalized equations (in terms of through and across variables) which describe the behavior of the generalized system that is analogous to the system shown in Exercise 6.1.

6.30. Write the generalized equations (in terms of through and across variables) which describe the behavior of the generalized system that is analogous to the system shown in Exercise 6.5.

6.31. Determine the transfer function $T(s)$ that relates the response $y(s)$ to the forcing function $f(s)$—in the Laplace domain—for the system shown in Exercise 6.1.

6.32. Determine the transfer function $T(s)$ that relates the response $x_1(s)$ to the forcing function $f(s)$—in the Laplace domain—for the system shown in Exercise 6.5.

6.33. Determine the transfer function $T(s)$ that relates the response $y(s)$ to the forcing function $f(s)$—in the Laplace domain—for the system shown in Exercise 6.17.

6.34. Determine the transfer function $T(s)$ that relates the response $y_1(s)$ to the forcing function $f(s)$—in the Laplace domain—for the system shown in Exercise 6.18.

6.35. Consider the system represented by the following block diagram:

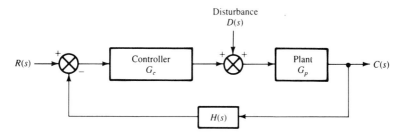

Disturbance
$D(s)$

(a) If the transfer functions are given by the expressions

$$G_c = \frac{2}{s}$$

$$G_p = \frac{36}{s + 10}$$

$$H = \frac{1}{s + 1}$$

$$R(s) = \frac{2}{s^2 + 4}$$

$$D(s) = \frac{3}{s + 5}$$

obtain the equivalent open-loop transfer function $T(s)$ for the system.
(b) Evaluate the system response for the case in which $D(s) = 0$.
(c) Evaluate the system response for the case in which $D(s) = 3/(s + 5)$.
(d) Evaluate the system response for the case in which $D(s) = 3$ (an impulse function).
(e) Compare the results of parts (a), (b), and (c); comment on the ability of the controller to manipulate the plant process properly for each type of disturbance $D(s)$.

6.36. Consider the management-production system shown in Figure 6.15. For the following values of the system parameters:

$$K_2 = 4, \quad K_4 = 1, \quad K_5 = 1, \quad \tau = 2$$

(a) Determine the system response $C(t)$ for $K_1 = 1$.
(b) Determine the system response $C(t)$ for $K_1 = 10$.
(c) Determine the system response $C(t)$ for

$$U_1(s) = 2 + \frac{1}{s}$$

(d) Determine the system resonse $C(t)$ for

$$G_1(s) = 2s + 1$$

(e) Compare the results of parts (a), (b), (c), and (d) and comment on your findings.

6.37. Write a computer program that will evaluate the system shown in Figure 6.15. The program should allow the user to specify the type of function $R(t)$ that will be applied to the system as its input (possibilities should include a parabolic or acceleration forcing function). Include complete documentation which explains both the use of the program and the objectives that can be achieved through its application.

6.38. Analyze the system shown in Figure 6.15, where the system parameters have the following values:

$$K_2 = 4, \quad K_4 = 1, \quad K_5 = 1, \quad \tau = 2$$

In particular:

(a) Determine the value of K_1 for which critical damping is achieved.

(b) For the value of K_1 determined in part (a), calculate the settling time for the system (based on a maximum variation of 2% about the equilibrium response for the system).

(c) Divide the value of K_1 which corresponds to critical damping, determined in part (a), by a factor equal to 2. Calculate the overshoot and settling time for the response (based on a 2% variation about the equilibrium response for the system) with this underdamped value of K_1. Compare this settling time to the value found in part (c).

(d) Multiply the value of K_1 which corresponds to critical damping, determined in part (a), by a factor equal to 2. Calculate the settling time for the response (based on a 2% variation about the equilibrium response for the system). Compare this settling time (for overdamped behavior) with the values found in parts (a) and (b).

6.39. A system has unity (negative) feedback with a feedforward transfer function $G(s)$ given by

$$G(s) = \frac{5}{s(s + 5)}$$

Determine the damping ratio ζ and the undamped natural frequency ω_n for this system.

6.40. A system has negative feedback with a feedforward transfer function $G(s)$ given by

$$G(s) = \frac{5}{s(s + 5)}$$

and a feedback transfer function $H(s)$ given by

$$H(s) = 2s + 1$$

Determine the damping ratio ζ and the undamped natural frequency ω_n for this system.

6.41. For the system shown below, determine the damping ratio ζ and the undamped natural frequency ω_n.

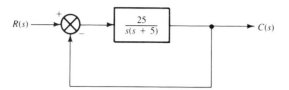

6.42. Consider the following block diagram representation of a speed control system for an automobile (Dorf, 1974; Holl, 1963).

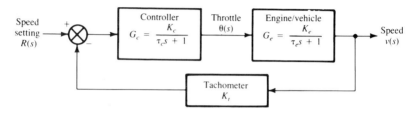

(a) Determine the equivalent open-loop transfer function $T(s)$ for this closed-loop system.

(b) Given $G_e(s)$ as shown in the diagram, relate the output speed $v(t)$ to the throttle (angular) position $\theta(t)$.

(c) Compare the behavior of the engine/vehicle component of the system, functionally described in $G_e(s)$, to that of the general first-order systems discussed in this chapter.

(d) Evaluate $v(t)$ for values of $\theta = 0°$, $15°$, $30°$, $45°$, and $60°$ in terms of the system parameters.

6.43. Determine the damping ratio ζ for the system described in Exercise 6.42 in terms of the system parameters. Also determine the undamped natural frequency ω_n for this system.

6.44. Determine the value of the controller gain K_c for the system described in Exercise 6.42 for which the damping ratio ζ equals 2, given that the other system parameters are as follows:

$$K_e = 200, \quad K_t = 1, \quad \tau_e = 30 \text{ s}, \quad \tau_c = 3 \text{ s}$$

6.45. Consider the following representation of a speed control system for an automobile to which is applied a disturbance $D(s)$ due to a road upgrade or downgrade (Dorf, 1974; Holl, 1963):

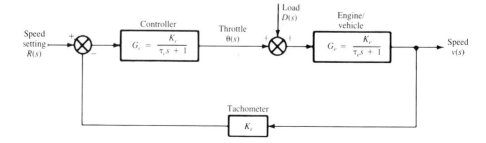

(a) Determine the equivalent open-loop transfer function $T(s)$ for this system.
(b) Determine $v(t)$ in terms of the system parameters, given that

$$R(s) = \frac{10}{s}, \qquad D(s) = \frac{0.1}{s}$$

(c) Determine $v(t)$ in terms of the system parameters, given that

$$R(s) = \frac{10}{s^2}, \qquad D(s) = \frac{0.1}{s}$$

(d) Determine $v(t)$ in terms of the system parameters, given that

$$R(s) = \frac{10}{s}, \qquad D(s) = 0$$

(e) Determine $v(t)$ in terms of the system parameters, given that

$$R(s) = \frac{10}{s^2}, \qquad D(s) = 0$$

Use the values for the system parameters given in Exercise 6.44, together with $K_c = 1$. Assume zero initial conditions.

6.46. Consider the following closed-loop block diagram description of an armature-controlled dc motor (adapted from Dorf, 1974):

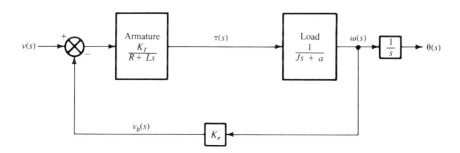

where $\tau(s)$, $\omega(s)$, and $\theta(s)$ represent the Laplace transforms of the torque $\tau(t)$, angular velocity $\omega(t)$, and angular displacement $\theta(t)$, respectively. In addition, the system parameters for this system include the armature voltage $v(t)$, the armature resistance R, the inductance L, the moment of inertia J of the applied load, the viscous damping coefficient a, the back emf $v_b(t)$, the torque constant K_T, and the voltage constant K_e.

(a) Determine the transfer function $T(s)$ for the equivalent open-loop block diagram of this system:

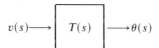

(b) Use an appropriate numerical approximation technique to determine the roots of the characteristic equation for this system, given that

$$K_e = 0.2 \text{ V/rad/s}$$

$$K_T = 28.0 \text{ oz-in./A}$$

$$R = 0.5 \text{ } \Omega$$

$$a = 0.1 \text{ in.-oz/rad/s}$$

$$J = 0.2 \text{ oz-in.-s}^2$$

$$L = 0.002 \text{ H}$$

and zero initial conditions.

(c) Evaluate the system response $\theta(t)$ to an applied voltage $v(t)$ equal to 25 V (with zero initial conditions).

6.47. Consider the following set of simultaneous differential equations which have been used to describe the relationship between a predator and its prey (Dorf, 1974):

$$\frac{dx_1}{dt} = ax_1 - \alpha x_2$$

$$\frac{dx_2}{dt} = bx_1 - \beta x_2$$

where $x_1 \equiv$ population of the prey

$x_2 \equiv$ population of predators

and where a, b, α, and β are constants for the system.

(a) Obtain a single second-order differential equation in which only x_1 appears as an unknown variable.

(b) Identify the damping ratio ζ and the undamped natural frequency ω_n, in terms of the system parameters, where $x_1(t)$ is defined as the system response.

(c) Given that

$$b = 1.5, \quad \alpha = 3.5, \quad \beta = 2.0$$

determine the maximum value of a for which the system developed in parts (a) and (b) will remain stable.

6.48. Consider the arms competition (or arms race) described in Section 6.4. Assume that all initial conditions are zero; furthermore, the system parameters have the following values:

$$a = 2.0, \quad b = 4.0, \quad h = 5.0, \quad f_1 = 50.0, \quad f_2 = 100.0$$

(a) Analyze this system and determine the maximum value for g if the system is to remain stable; assume that the system response of interest is x_1.

(b) Determine the system response $x_1(t)$, given that the value of g is equal to $0.8g_{max}$, where g_{max} represents the maximum value of g [determined in part (a)] for which the system will remain stable.

(c) Determine the system response $x_2(t)$, given that the value of g is equal to $0.8g_{max}$.

7

ACCURACY, STEADY-STATE ERROR, AND CONTROL ACTIONS

It is quite a three pipe problem, and I beg that you won't speak to me for fifty minutes.

Sherlock Holmes, The Red-Headed League *by Sir Arthur Conan Doyle*

7.1 OBJECTIVES

Upon completion of this chapter, the reader should be able to:

- Evaluate the steady-state error (if any) within a system for a given forcing function.
- Identify a system's Type number.
- Determine the steady-state error that is to be expected upon application of a step input, a ramp input, or a parabolic (acceleration) input to a given system type.
- Identify and apply the following types of control actions:
  (a) Two-position or on-off control
  (b) Proportional control
  (c) Integral control
  (d) Proportional-plus-integral control
  (e) Proportional-plus-derivative control
  (f) Proportional-plus-derivative-plus-integral control
- Explain and apply the basic principle for generating the desired control action via the introduction of the appropriate elements in the feedback path of a closed loop.

179

7.2 ACCURACY AND STEADY-STATE ERROR

7.2.1 Steady-State Error and System Type

Our two major goals in control efforts, as noted in Chapter 1, are *stability* and *accuracy*. We have discussed the goal of stability at some length in preceding chapters and will return to this goal in Chapter 8; we now direct our attention toward our other primary goal—accuracy.

The degree to which a system does *not* achieve the goal of accuracy *after* the system has achieved equilibrium (or its final steady state) can be measured in terms of the steady-state error e_{ss}, defined by the expression

$$e_{ss} = R(t) - C(t) \qquad (7.1)$$

where $R(t)$ represents the system's reference input (a measure of the desired output response) and $C(t)$ is the actual steady-state response of the system. For a closed-loop system with negative feedback (as shown in Figure 7.1), the actuating signal E is given by

$$E = \frac{R(s)}{1 + KG(s)H(s)} \qquad (7.2)$$

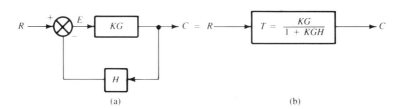

Figure 7.1   Standard closed-loop system. (a) closed-loop diagram (b) equivalent open-loop form.

(to a first-order approximation). Note that E is *not* identical to e_{ss} [compare equations (5.11) and (7.1)]; however, E does measure the system performance in terms of the accuracy achieved by the system. If the system had unity feedback (wherein $H = 1$), then

$$E = R(s) - C(s)$$
$$= \frac{R(s)}{1 + KG(s)} \qquad (7.3)$$

and E does represent e_{ss} in the Laplace domain. For simplicity, we will assume unity feedback in the following discussions.

$KG(s)$ represents the transfer function of the process under consideration; it may be represented in a general form as

$$KG(s) = \frac{K(s - z_1)(s - z_2) \cdots (s - z_i)}{s^n(s - p_1)(s - p_2) \cdots (s - p_l)} \tag{7.4}$$

[where there are i values $(z_1, z_2, \ldots, z_i)$ of s which are *zeros* of $KG(s)$, that is, values of s for which $KG(s)$ equals zero, and where there are l values $(p_1, p_2, \ldots, p_l)$ of s which are *poles* of $KG(s)$, that is, values of s for which $KG(s)$ becomes infinite; in addition, there are n poles located at the origin of the complex s-plane]. If we wrote the equivalent open-loop transfer function $T(s)$ for this system, we would have

$$T(s) = \frac{KG(s)}{1 + KG(s)} \tag{7.5}$$

in accordance with equation (5.16). Given that operation on the input $R(s)$ by $T(s)$ will result in the output $C(s)$, we have

$$C(s) = T(s)R(s) \tag{7.6}$$

Recall that if the differential equation

$$(a_n D^n + a_{n-1} D^{n-1} + \cdots + a_1 D + a_0)C(s) = R(s) \tag{7.7}$$

also describes the behavior of the system, then the characteristic equation for the system is

$$a_n D^n + a_{n-1} D^{n-1} + \cdots + a_1 D + a_0 = 0 \tag{7.8}$$

One can then identify corresponding relations, resulting in (for zero initial conditions)

$$\frac{1}{T(s)} = a_n s^n + a_{n-1} s^{n-1} + \cdots a_1 s + a_0 \tag{7.9}$$

or, for the characteristic equation,

$$1 + KG(s) = 0 \tag{7.10}$$

As a result, we then obtain

$$1 + KG(s) = 1 + \frac{K(s - z_1)(s - z_2) \cdots (s - z_i)}{s^n(s - p_1) \cdots (s - p_l)} \tag{7.11}$$

or

$$s^n(s - p_1)(s - p_2) \cdots (s - p_l) + K(s - z_1)(s - z_2) \cdots (s - z_i) = 0 \quad (7.12)$$

Systems are classified by *type number* in accordance with the number n of poles [of $KG(s)$] located at the origin of the s-plane. For example, the values $n = 0, 1, 2$ correspond to type 0, type 1, and type 2 systems. Notice that the order of the system is *not* necessarily identical to the system's type number, where the order of the system corresponds to the number of roots of the characteristic equation. (We will demonstrate in a forthcoming section of this chapter that the system's type can be modified through the introduction of a particular kind of control mechanism known as integral control.)

In addition, we now introduce the following quantities, which will prove to be very convenient in our forthcoming work:

$$K_0 = \lim_{s \to 0} KG(s) \equiv \text{positional error constant}$$

$$K_v = \lim_{s \to 0} sKG(s) \equiv \text{velocity error constant}$$

$$K_a = \lim_{s \to 0} s^2 KG(s) \equiv \text{acceleration error constant}$$

(following the terminology used in Dorf, 1974). With these quantities, we may consider steady-state error for three distinct types of forcing functions: a step input, a ramp input, and a parabolic (in time t) input.

7.2.2 Steady-State Error for a Step Input

If the input or forcing function $R(t)$ is a step input, that is,

$$R(t) = A \qquad \text{for } t \geq 0 \tag{7.13}$$

or

$$R(s) = \frac{A}{s} \tag{7.14}$$

(where A is a constant), the steady-state error e_{ss} is given by (for a unity feedback system in which E equals e_{ss})

$$e_{ss}(s) = \frac{R(s)}{1 + KG(s)} \tag{7.15}$$

Recall that the final-value theorem allows us to determine the value of a function of time as t approaches infinity, in accordance with the relation

$$\lim_{t \to \infty} f(t) = \lim_{s \to 0} sF(s)$$

The steady-state error e_{ss} is that error which exists in the system after the transient response has vanished (for a stable system) and the system has achieved equilibrium (i.e., the error that exists as time t becomes large). Therefore, applying the final-value theorem to the expression (7.15), we obtain

$$e_{ss} = \lim_{s \to 0} se_{ss}(s)$$

$$= \text{steady-state error in the time domain} \tag{7.16}$$

For the step input,

$$e_{ss} = \lim_{s \to 0} \frac{s(A/s)}{1 + KG(s)} \qquad (7.17)$$

or, since A is a constant,

$$e_{ss} = \frac{A}{1 + \lim_{s \to 0} KG(s)} \qquad (7.18)$$

Recalling the definition of the positional error constant K_0, we have

$$e_{ss} = \frac{A}{1 + K_0} \qquad (7.19)$$

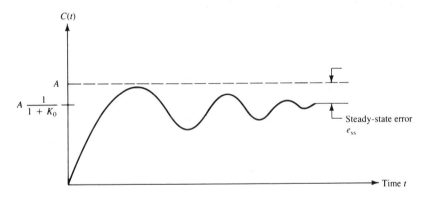

Figure 7.2   Steady-state error for a step input.

Figure 7.2 shows the significance of this result: As K_0 increases, the steady-state error decreases so that the system approaches the desired behavior in its equilibrium state. If one required an accuracy level of 99% [i.e., a steady-state error of 1% or less of the desired response $R(t)$], then

$$e_{ss} = 0.01A$$

$$= \frac{A}{1 + K_0}$$

or

$$K_0 = 99$$

For a type 0 *system*, there are no poles of $KG(s)$ at the origin of the complex s-plane. Given the general expression (7.4) for $KG(s)$, we have

$$e_{ss} = \frac{A}{1 + \lim\limits_{s \to 0} \dfrac{K(s - z_1)(s - z_2) \cdots (s - z_i)}{s^0(s - p_1) \cdots (s - p_l)}}$$

$$= \frac{A}{1 + \dfrac{K(-z_1)(-z_2) \cdots (-z_i)}{(-p_1)(-p_2) \cdots (-p_l)}} \tag{7.20}$$

(e_{ss} decreases as K increases). Notice that a finite error will exist if a step input is applied to a type 0 system, since K must necessarily be finite.

For a type 1 *system*, there is a single pole of $KG(s)$ at the origin of the s-plane. We then have

$$e_{ss} = \frac{A}{1 + \lim\limits_{s \to 0} \dfrac{K(s - z_1)(s - z_2) \cdots (s - z_i)}{s^1(s - p) \cdots (s - p_l)}}$$

$$= \frac{A}{1 + \infty} \to 0 \tag{7.21}$$

In other words, a step input applied to a type 1 system will result in a system response without steady-state error.

For a type $n \geq 1$ *system*, e_{ss} is equal to zero, as can be shown via an argument similar to that above for type 1 systems.

7.2.3 Steady-State Error for a Ramp Input

If the forcing function $R(t)$ applied to a system is a ramp input (proportional to time t):

$$R(t) = vt \tag{7.22}$$

so that

$$R(s) = \frac{v}{s^2} \tag{7.23}$$

(where v is a constant of proportionality), the steady-state error e_{ss} is given by

$$e_{ss} = \frac{R(s)}{1 + KG(s)}$$

$$= \frac{v}{s^2[1 + KG(s)]} \tag{7.24}$$

Given the general expression (7.4) for the feedforward transfer function $KG(s)$, we then have

$$e_{ss} = \lim_{s \to 0} s e_{ss}(s)$$

$$= \lim_{s \to 0} \frac{v}{s[1 + KG(s)]}$$

$$= \lim_{s \to 0} \frac{v}{s KG(s)} \tag{7.25}$$

$$= \frac{v}{K_v}$$

where K_v is the velocity error constant introduced earlier.

For a type 0 system, the steady-state error e_{ss} is

$$e_{ss} = \lim_{s \to 0} \frac{v \cdot s^0 (s - p_1) \cdots (s - p_l)}{K \cdot s^1 (s - z_1) \cdots (s - z_i)} \tag{7.26}$$

$$\to \infty$$

For a type 1 system, the steady-state error is

$$e_{ss} = \lim_{s \to 0} \frac{v \cdot s^1 (s - p_1) \cdots (s - p_l)}{K \cdot s^1 (s - z_1) \cdots (s - z_i)}$$

$$= \frac{v(-p_1) \cdots (-p_l)}{K(-z_1) \cdots (-z_i)} \tag{7.27}$$

(a finite value) or

$$e_{ss} = \frac{v}{K_v} \tag{7.28}$$

(see Figure 7.3).

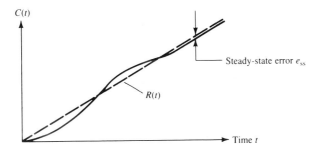

Figure 7.3   Steady-state error for a ramp input.

For a type n ≥ 2 system, the steady-state error e_{ss} vanishes, that is,

$$e_{ss} = 0 \tag{7.29}$$

7.2.4 Steady-State Error for a Parabolic (Acceleration) Input

If the forcing function $R(t)$ applied to a system is an acceleration input (i.e., proportional to t^2 or parabolic in t):

$$R(t) = \tfrac{1}{2}at^2 \tag{7.30}$$

(where a is a constant), so that

$$R(s) = \frac{a}{s^3} \tag{7.31}$$

then

$$e_{ss} = \lim_{s \to 0} \frac{sR(s)}{1 + KG(s)}$$

$$= \lim_{s \to 0} \frac{a}{s^2 + s^2 KG(s)}$$

$$= \frac{a}{\lim_{s \to 0}[s^2 KG(s)]} \tag{7.32}$$

$$= \frac{a}{K_a}$$

where K_a is the acceleration error constant.

Given the general expression (7.4) for $KG(s)$, we may then consider the following types of systems to which a ramp input could be applied and the corresponding steady-state error to be expected.

For a type 0 system, the acceleration error constant K_a is

$$K_a = \lim_{s \to 0} s^2 KG(s) = \lim_{s \to 0} \frac{s^2 K(s - z_1) \cdots (s - z_i)}{s^0(s - p_1) \cdots (s - p_l)}$$

$$\to 0$$

so that

$$e_{ss} \to \infty \tag{7.33}$$

For a type 1 system, we again have

$$K_a \to 0$$

so that

$$e_{ss} \to \infty \tag{7.34}$$

For a type 2 *system*, the acceleration error constant K_a is

$$K_a = \lim_{s \to 0} \frac{s^2 K(s - z_1) \cdots (s - z_i)}{s^2(s - p_1) \cdots (s - p_l)}$$
$$= \frac{K(-z_1)(-z_2) \cdots (-z_i)}{(-p_1)(-p_2) \cdots (-p_l)} \tag{7.35}$$

(a finite value) or

$$e_{ss} \to \frac{a}{K_a} \tag{7.36}$$

A finite steady-state error is then predicted in the event that an acceleration input (parabolic in t) is applied to a type 2 system (see Figure 7.4); as K_a is increased, e_{ss} is decreased.

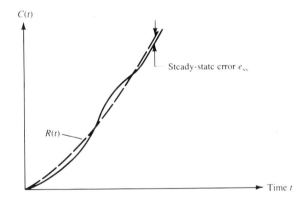

Figure 7.4   Steady-state error for an acceleration (parabolic) input.

For a type $n > 2$ *system*, K_a approaches infinity so that

$$e_{ss} \to 0 \tag{7.37}$$

7.2.5 Summary

The results of the previous three sections are summarized in Table 7.1. Notice that as one increases the system type number, the steady-state error e_{ss} can be reduced to zero. One increases type number by increasing the number of poles of $KG(s)$ located at the origin of the s-plane; such an increase in the number of poles at the origin corresponds to an increase in the number of integral control actions within the system's closed loop. As a result, we

TABLE 7.1   **Steady-State Error e_{ss} Values**

System Type	Input (Forcing Function)		
	$R(t) = A$ (Step Input)	$R(t) = vt$ (Ramp Input)	$R(t) = 0.5at^2$ (Acceleration Input)
0	$A/(1 + K_0)$	∞	∞
1	0	v/K_v	∞
2	0	0	a/K_a
3	0	0	0
$\vdots$	$\vdots$	$\vdots$	$\vdots$

now consider common control actions (including integral control) that can be applied to a system with feedback.

7.3 CONTROL ACTIONS

7.3.1 General Control Action

Figure 7.5 presents the block diagram of a system in which a process is monitored and controlled via feedback. The actuating signal $E(s)$—in the Laplace (s) domain—is described mathematically by

$$E(s) = R(s) - C(s) \tag{7.38}$$

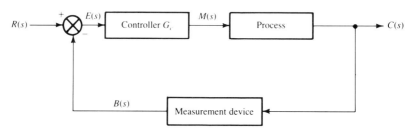

Figure 7.5   Example of a closed-loop system.

in accordance with the difference operator that produces $E(s)$ as its output signal. $R(s)$ is the Laplace transform of the system's reference input and $C(s)$ is the Laplace transform of the system's controlled output response. $M(s)$ represents the Laplace transform of the manipulated signal from the controller element and $B(s)$ is the feedback signal to the difference operator in the s-domain (see Schwarzenbach and Gill, 1978).

If the controller has a transfer function $G_c(s)$, the manipulated signal $M(s)$ is related to $E(s)$ according to the relation

$$M(s) = G_c(s)E(s) \qquad (7.39)$$

The transfer function $G_c(s)$ describes the type of control action that is used in the system. We now consider several different types of control actions which are commonly used in automatic control systems. Equipment instrumentation (for example, transducers, actuators and sensors) used to incorporate these control actions within mechanical, electrical, and other types of systems are described in references such as Hunter (1978), Lenk (1980), and Lenk (1984).

7.3.2 Two-Position or On-Off Control

Two-position or on-off control can be achieved with the use of a controller that produces only two manipulated (correction) signals: M_1 and M_2. (An electrical relay can be used as a two-position control element in many situations.) A block diagram representation of a two-position controller is shown in Figure 7.6a, where, depending on the value of the actuating signal $E(s)$, the controller's output signal $M(s)$ equals M_1 or M_2. M_2 is frequently equal to either $-M_1$ or zero, as in the case of a relay in which there is an output current ($M = M_1$) or there is no current ($M = 0$, i.e., an open circuit).

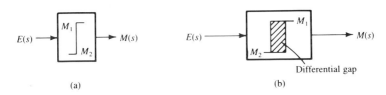

(a)   (b)

Figure 7.6   Two-position or on-off controller.

A differential gap or neutral zone (see Figure 7.6b) is often intentionally included in the design of a two-position controller in order to prevent frequent (on-off) switching of the element that accelerates wear (Ogata, 1970; Palm, 1983). For example, a thermostatic control of a heating system may include a neutral zone of approximately 2°C about the set temperature or operating point. A differential gap may also result from frictional effects or other physical actions which have been neglected in the system's model or design.

7.3.3 Proportional Control

As its name implies, a proportional controller produces a modulated signal $M(s)$ which is proportional to the actuating (error) signal $E(s)$, that is,

$$M(s) = K_p E(s) \qquad (7.40)$$

where K_p is the constant of proportionality [known as the *proportional sensitivity* or the *gain* (see Ogata, 1970)]. In block diagram notation, the proportional controller is represented as follows:

$$E(s) \longrightarrow \boxed{\;\;K_p\;\;} \longrightarrow M(s)$$

that is, the transfer function $G_c(s)$ of the proportional controller is simply equal to K_p.

In an electronic system, a proportional controller is essentially an amplifier with an adjustable gain K_p. In liquid-level systems, a float-control is often used to achieve proportional control: The float rises as the liquid level increases, thereby causing a proportional decrease in the liquid flow rate into the vessel through an associated movement of a valve. As a result, a relatively constant liquid level is maintained.

7.3.4 Integral Control

With integral control action, the rate of change dM/dt in the modulated signal is proportional to the actuating signal, that is,

$$\frac{dM}{dt} = K_i E(t) \tag{7.41}$$

where K_i (known as the *integral gain*) is an adjustable constant. If we then integrate this expression for $M(t)$, we obtain

$$M(t) = K_i \int_0^t E(t)\,dt \tag{7.42}$$

(hence the name for this control action). Transforming equation (7.41) to the complex Laplace s-domain, we have (assuming zero initial conditions)

$$sM(s) = K_i E(s) \tag{7.43a}$$

or

$$M(s) = \frac{K_i}{s} E(s) \tag{7.43b}$$

Since

$$M(s) = G_c(s)E(s)$$

where $G_c(s)$ is the transfer function of the controller, we then know that

$$G_c(s) = \frac{K_i}{s} \tag{7.44}$$

for an integral control element. In block diagram notation, such an element is represented as follows:

$$E(s) \longrightarrow \boxed{\dfrac{K_i}{s}} \longrightarrow M(s)$$

Notice that the rate at which correction is sought increases with the error [of which $E(t)$ is a measure] contained within the system. Furthermore, the controller continues to operate until the error is zero (see Schwarzenbach and Gill, 1978).

Recall that, for a given forcing function, steady-state error e_{ss} could be reduced within a system if one increased the number n of open-loop poles located at the origin of the s-plane (i.e., if one increased the type number of the system). Such an increase in the type number of the system can be achieved by inserting integral control elements into the loop of the system since each such element results in a multiplication of the product of component transfer functions (KGH) by a factor K_i/s. However, as we shall demonstrate, the addition of integral control elements within the system's closed loop can lead to instability, so that we may be limited in our use of such elements in order to minimize steady-state error.

7.3.5 Proportional-Plus-Integral Control

A combination of proportional and integral control actions can be achieved in accordance with the block diagram representation

$$E(s) \longrightarrow \boxed{K_p\left(1 + \dfrac{1}{T_i s}\right)} \longrightarrow M(s)$$

where T_i is an adjustable constant (known as the *integral time*) which allows one to modify the integral control action (see Ogata, 1971). Mathematically, the modulated signal $M(s)$ is related to the actuating signal according to

$$\begin{aligned} M(s) &= G_c(s)E(s) \\[2mm] &= K_p\left(1 + \frac{1}{T_i s}\right)E(s) \end{aligned} \tag{7.45}$$

or, in the time domain,

$$M(t) = K_p E(t) + \frac{K_p}{T_i} \int_0^t E(t)\, dt \qquad (7.46)$$

Figure 7.7 shows the modulated signal $m(t)$ for proportional, integral, and proportional-plus-integral control actions, given a unit step input actuating signal $E(t)$ (Ogata, 1970).

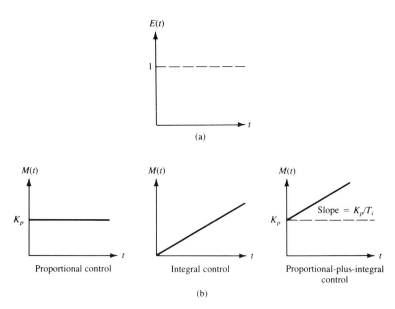

Figure 7.7   (a) Step-input actuating signal $E(t)$; (b) various modulated signals $M(t)$ for different types of control actions, given the step input $E(t)$. (Adapted from Ogata, 1971).

As we will demonstrate in a forthcoming section of this chapter, the addition of integral control action results in the elimination of steady-state error in type 0 systems in which only proportional control was initially used.

7.3.6 Proportional-Plus-Derivative Control

Derivative control action results if the modulated signal $M(t)$ is proportional to the rate of change dE/dt of the actuating signal; however, such derivative control is *never* used without another type of control action since it will produce a signal only if the error is varying (Ogata, 1971; Schwarzenbach and Gill, 1978). During nontransient periods in which the error is constant (but not zero), no correction signal $M(t)$ would be generated by a controller that is only capable of derivative control action. Proportional-plus-derivative control action can be mathematically described by the expression

$$M(t) = K_p \left[E(t) + T_d \frac{dE}{dt} \right] \qquad (7.47)$$

where T_d is the derivative time. [One could also define a derivative gain K_d such that, for the derivative control action only,

$$M(t) = K_d \frac{dE}{dt} \qquad (7.48)$$

(see Palm, 1983).] Upon Laplace transformation (with zero initial conditions), expression (7.47) becomes

$$\begin{aligned} M(s) &= (K_p + K_p T_d s) E(s) \\ &= G_c(s) E(s) \end{aligned} \qquad (7.49)$$

or, for proportional-plus-derivative control, the corresponding transfer function $G_c(s)$ for the controller is given by

$$G_c(s) = K_p + K_p T_d s \qquad (7.50)$$

Derivative control action provides an anticipatory effect for the system (Schwarzenbach and Gill, 1978). A large rate dE/dt indicates that the system may overshoot the desired value for the output $C(t)$ by a wide margin; however, $M(t)$ also will be large if derivative control is used, thereby providing correction for the (anticipated) error. As a result, derivative control can be used to increase the stability of a system (Ogata, 1978).

In addition, for the case of *proportional-plus-derivative* control, $M(t)$ is composed of two terms, one of which is proportional to the actuating signal $E(t)$ and one of which is proportional to the rate of change of $E(t)$. Consider Figure 7.8, in which the actual output $C(t)$ is shown as a function of time t

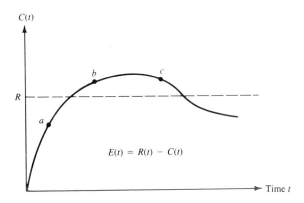

Figure 7.8   Derivative control action at different times during the system response.

relative to a desired constant output signal $R(t)$ [where $R(t)$ is the reference input]. For unity feedback [in which $H(s)$ equals 1], we know that the actuating signal $E(t)$ is identical to the error $e(t)$ within the system:

$$E(t) = e(t)$$
$$= R(t) - C(t)$$

In addition, noting that $R(t)$ is a constant for the case under consideration, we obtain

$$\frac{dE}{dt} = \frac{d(R - C)}{dt}$$
$$= -\frac{dC}{dt} \tag{7.51}$$

When the system is in the transient state a shown in Figure 7.8, the signal $E(t)$ is positive in sign, whereas dE/dt is negative (note that dC/dt is a positive quantity at state a); as a result, $M(t)$ is reduced by an amount dE/dt (due to the derivative control portion of the controller) from the value $K_p E(t)$ that would be produced if only proportional control were used. Similarly, when the system is in state c, $E(t)$ and dE/dt are opposite in sign [$E(t)$ is negative and dE/dt is positive], so that once again dE/dt tends to reduce the correction signal $M(t)$. However, if the system is in state b, both $E(t)$ and dE/dt are identical in sign, so that dE/dt augments the proportional control signal $K_p E(t)$. We may generalize these observations by stating that (1) the derivative control action increases the correction signal $M(t)$ if the output $C(t)$ is moving away from its desired value $R(t)$, and (2) the derivative control action decreases $M(t)$ if $C(t)$ is moving toward the desired value. Finally, if dE/dt is equal to zero, the derivative control action has no effect on the system. [For a detailed discussion of these observations, see Schwarzenbach and Gill (1978).]

7.3.7 Proportional-Plus-Derivative-Plus-Integral Control

If we combine proportional, derivative, and integral control actions, we obtain a controller that can be represented by the following block diagram:

$$E(s) \longrightarrow \boxed{\; K_p\left(1 + T_d s + \frac{1}{T_i s}\right) \;} \longrightarrow M(s)$$

where the transfer function $G_c(s)$ for this control element is given by

$$G_c(s) = K_p \left(1 + T_d s + \frac{1}{T_i s} \right) \tag{7.52}$$

As will be seen in the next section, the addition of derivative control action to proportional-plus-integral control (discussed in Section 7.3.4) results in a dampening of any oscillations in the output $C(t)$ of the system.

7.4 APPLICATION OF VARIOUS CONTROL ACTIONS

7.4.1 Example Process

We now apply various control actions to the second-order system shown in Figure 7.9. (Note that, for simplicity, this system has unity feedback; the analysis would become more cumbersome algebraically if nonunity feedback were considered without providing us with the benefit of additional clarity.)

We will assume that the process is second-order with a transfer function $G(s)$ given by

$$G(s) = \frac{K}{s^2 + ps + q} \tag{7.53}$$

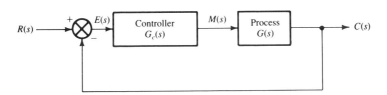

Figure 7.9   Example system.

where K, p, and q are constants which are specific to the process under consideration. Furthermore, for simplicity, we will assume that the forcing function $R(t)$ is a step input, that is,

$$R(t) = A \qquad \text{for } t \geq 0$$
$$= \text{constant} \tag{7.54}$$

so that

$$R(s) = \frac{A}{s} \tag{7.55}$$

7.4.2 Application of Proportional Control

For the case of proportional control, in which

$$G_c(s) = K_p \tag{7.56}$$

we know that the system may be written in the equivalent open-loop block diagram form as

$$R(s) \longrightarrow \boxed{T_p(s)} \longrightarrow C(s)$$

where the system's transfer function $T_p(s)$ is given by

$$T_p = \frac{K_p G(s)}{1 + K_p G(s)} \tag{7.57}$$

[where $G(s)$ is given by equation (7.53)]. Therefore,

$$
\begin{aligned}
C(s) &= T_p(s) R(s) \\
&= \frac{K_p G(s)}{1 + K_p G(s)} \frac{A}{s} \\
&= \frac{A K_p K}{s(s^2 + ps + q + KK_p)}
\end{aligned}
\tag{7.58}
$$

or, upon expansion in partial fractions,

$$C(s) = \frac{C_1}{s} + \frac{B_1 s + B_2}{s^2 + ps + q + KK_p} \tag{7.59}$$

where C_1, B_1, and B_2 are constants to be determined. If we assume that p, q, K, and K_p are such that the roots (r_1 and r_2) of the quadratic factor form a complex conjugate pair, we then have

$$r_1, r_2 = a \pm jb \tag{7.60}$$

where

$$a = -\frac{p}{2} \tag{7.61a}$$

$$b = \tfrac{1}{2}(4q + 4KK_p - p^2)^{1/2} \tag{7.61b}$$

As a result, we may then obtain

$$C(t) = C_1 + C_2 e^{at} \sin(bt + \phi) \tag{7.62}$$

(where C_2 and ϕ are to be determined). For a stable system (in which $a < 0$), we obtain

$$C(t) \to C_1 = \frac{AKK_p}{q + KK_p}$$

$$< A$$

(7.63)

as time t approaches infinity. As a result, the system then has a steady-state error given by

$$e_{ss} = R(t) - C(t)$$

$$= A - \frac{AKK_p}{q + KK_p}$$

$$= A\frac{q}{q + KK_p}$$

(7.64)

(see Figure 7.10). As the proportional gain K_p is increased, e_{ss} decreases; however, for a finite K_p, we must expect a nonvanishing steady-state error e_{ss}. Furthermore, as we will demonstrate in Chapter 8, the system can become unstable as K_p is increased. To eliminate e_{ss}, we now introduce integral control action.

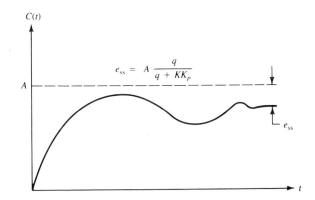

Figure 7.10   Response of example system with proportional control action.

7.4.3 Application of Proportional-Plus-Integral Control

If we apply proportional-plus-integral control action to the system shown in Figure 7.11, we effectively change the system from type 0 to type 1, since the equivalent open-loop transfer function $T_{PI}(s)$ for the system (Figure 7.11)

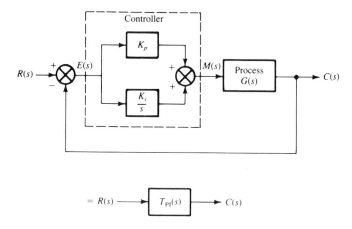

Figure 7.11   Example system with proportional-plus-integral control action.

is then

$$T_{PI}(s) = \frac{K(K_p s + K_i)}{s(s^2 + ps + q + KK_p) + K_i K} \tag{7.65}$$

so that, for $R(s) = A/s$,

$$
\begin{aligned}
C(s) &= T_{PI}(s)R(s) \\[2ex]
&= \frac{AKK_p s + AKK_i}{s(s^3 + ps^2 + (q + KK_p)s + K_i K}
\end{aligned}
\tag{7.66}
$$

or, upon application of the final-value theorem,

$$
\begin{aligned}
C(t \to \infty) &= \lim_{s \to 0} sC(s) \\[2ex]
&= A
\end{aligned}
\tag{7.67}
$$

so that the steady-state error vanishes, as desired. Figure 7.12 shows the response of this system as a function of time t. Notice that the output may significantly overshoot the desired value A during the transient portion of the response; to minimize such overshoot, we now include derivative control action in this system.

7.4.4 Application of Proportional-Plus-Derivative-Plus-Integral Control

Proportional-plus-derivative-plus-integral control action, as shown in Figure 7.13, results in an equivalent open-loop transfer function $T_{PDI}(s)$ given

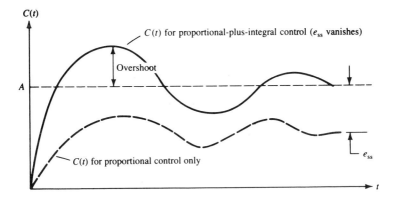

Figure 7.12 Response of example system with proportional-plus-integral control action.

by

$$T_{\text{PDI}}(s) = \frac{KK_d s^2 + KK_p s + KK_i}{s^3 + [KK_d + p]s^2 + [KK_p + q]s + KK_i} \tag{7.68}$$

For a step input $R(s) = A/s$, we then obtain

$$C(s) = T_{\text{PDI}}(s)R(s)$$

$$= \frac{A(KK_d s^2 + KK_p s + KK_i)}{s[s^3 + (KK_d + p)s^2 + (KK_p + q)s + KK_i]}$$

or, through application of the final-value theorem,

$$C(t \to \infty) = \lim_{s \to 0} sC(s)$$

$$= A$$

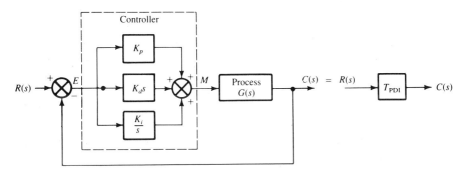

Figure 7.13 Example system with proportional-plus-derivative-plus-integral control action.

as we obtained for the case of proportional-plus-integral control. However, upon comparison of the denominators of $T_{\text{PI}}(s)$ and $T_{\text{PDI}}(s)$, where

$$T_{\text{PI}}(s) = \frac{K(K_p s + K_i)}{s^3 + ps^2 + (q + KK_p)s + KK_i} \tag{7.65}$$

we notice that the only difference is the additional term $KK_d s^2$ in the case of PDI control. Since the denominator of the equivalent open-loop transfer function T of a closed-loop system corresponds to the characteristic equation when set equal to zero, this additional term $KK_d s^2$ corresponds to increased damping of the oscillatory transient behavior of $C(t)$, as shown in Figure 7.14; in other words, the damping coefficient p of the second-order process under consideration [see equation (7.53)] is increased by an amount equal to KK_d.

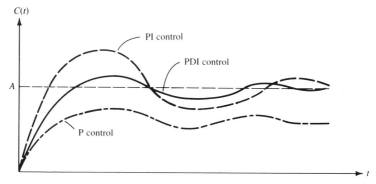

Figure 7.14 Response of example system with proportional-plus-derivative-plus-integral control action.

7.5 BASIC PRINCIPLE FOR CONTROL GENERATION

In conclusion, we note that—for physical systems with *nonunity* feedback, as shown in Figure 7.15—it is often true that the absolute value $|G(s)H(s)|$ of the product of the transfer functions for components within the system's closed loop is such that

$$|G(s)H(s)| \gg 1$$

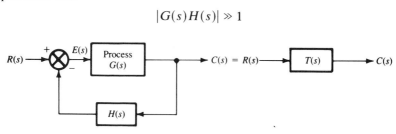

Figure 7.15 Standard control system; $T(s) \doteq H^{-1}(s)$ if $|G(s)H(s)| \gg 1$.

As a result, the equivalent open-loop transfer function $T(s)$ can be approximated by

$$T(s) = \frac{G(s)}{1 + G(s)H(s)}$$

$$\approx \frac{G(s)}{G(s)H(s)}$$

$$= \frac{1}{H(s)}$$

Since

$$C(s) = T(s)R(s)$$

$$\approx \frac{R(s)}{H(s)}$$

one can then obtain the desired control action for the system by inserting a component into the feedback path of the loop where the transfer function of this component is the inverse of the desired control action. For example, if PI control is desired, one inserts a component with the transfer function

$$H(s) = \frac{1}{K_p[1 + (1/T_i s)]} \tag{7.69}$$

into the feedback path. This approach to the design of a control system is often used (Ogata, 1970; Palm, 1983).

7.6 REVIEW

In summary, we have reviewed the following topics, facts, relationships, or concepts in this chapter.

- Steady-state error e_{ss}, as its name implies, is the error within the system after equilibrium has been achieved. It is expressed mathematically in the form

$$e_{ss} = R(t) - C(t) \tag{7.1}$$

  where $R(t)$ is the reference input and $C(t)$ is the steady-state response of the system.

- For a closed-loop system, the actuating signal $E(s)$ can be expressed as

$$E(s) = \frac{R(s)}{1 + KG(s)H(s)} \tag{7.2}$$

where K is a constant, $G(s)$ is the product of the feedforward transfer functions, and $H(s)$ is the product of the feedback transfer functions contained in the loop. For unity feedback, $H(s)$ equals unity and $e_{ss}(s)$ is identical to $E(s)$.

- The final-value theorem allows us to calculate the steady-state error in accordance with the expression

$$e_{ss} = \lim_{s \to 0} [s e_{ss}(s)] \tag{7.16}$$

- Systems are classified by type number in accordance with the number (n) of poles of $KG(s)H(s)$ located at the origin of the complex s-plane. Systems of type $0, 1, 2, \ldots, j$ correspond to $n = 0, 1, 2, \ldots, j$. The order of the system is *not* necessarily identical to the system's type number.

- Error constants were introduced as follows:

$$K_0 = \lim_{s \to 0} KG(s) = \text{positional error constant}$$

$$K_v = \lim_{s \to 0} sKG(s) = \text{velocity error constant}$$

$$K_a = \lim_{s \to 0} s^2 KG(s) = \text{acceleration error constant}$$

- Table 7.1 was developed, which indicates that the steady-state e_{ss} is dependent on both the system design (type) and the specific forcing function applied to the system.

- Control actions were introduced, including
  (a) Two-position or on-off control
  (b) Proportional control with a transfer function

$$G_c(s) = K_p$$

  (c) Integral control with a transfer function

$$G_c(s) = \frac{K_i}{s}$$

  (d) Proportional-plus-derivative control with a transfer function

$$G_c(s) = K_p(1 + T_d s)$$

  (e) Proportional-plus-integral control with a transfer function

$$G_c(s) = K_p\left(1 + \frac{1}{T_i s}\right)$$

(f) Proportional-plus-derivative-plus-integral control with a transfer function

$$G_c(s) = K_p\left(1 + T_d s + \frac{1}{T_i s}\right)$$

- The application of each control action identified above was discussed, together with the effects of such control actions on system response.
- Finally, the basic principle for generating a desired control action was introduced, where it was assumed that $|G(s)H(s)| \gg 1$, so that one may simply insert an element in the feedback path of the closed loop with a transfer function $H(s)$ which is the inverse of the desired control action.

EXERCISES

7.1. If a system's equivalent open-loop transfer function $T(s)$ is given by

$$T(s) = \frac{s^2 + 3s + 2}{1 + (s^2 + 3s + 2)}$$

determine:
(a) The steady-state error for the system for a step input $R(t) = 10$.
(b) The steady-state error for the system for a ramp input $R(t) = 3t$.

7.2. A system with unity feedback has a feedforward transfer function given by

$$KG(s) = \frac{s^2 + 2s + 1}{s^2(s + 1)(s + 4)^2}$$

(a) Determine the steady-state error for the system on application of a step input $R(t) = 5$.
(b) Determine the steady-state error for the system on application of an acceleration input $R(t) = 6t^2$.
(c) Identify the system type.

7.3. Consider the system represented by the following block diagram:

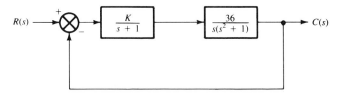

(a) Determine the system type.
(b) Determine the steady-state error for the system if

$$R(s) = \frac{12}{s} \qquad K = 0.1$$

(c) Determine the steady-state error for the system if

$$R(s) = \frac{12}{s} \qquad K = 10$$

(d) Determine the steady-state error for the system if

$$R(s) = \frac{20}{s^2} \qquad K = 0.1$$

7.4. Consider the system represented by the following block diagram:

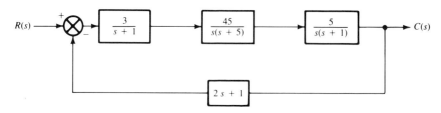

(a) Determine the steady-state error if $R(t) = 3t^2$.
(b) Determine the steady-state error if $R(t) = 12$.
(c) Identify the system type.

7.5. Given a system with unity feedback and feedforward transfer functions

$$G_1(s) = \frac{20}{s(s+1)} \qquad G_2(s) = \frac{K}{s(s^2 + 3s + 2)}$$

where K is a constant to be determined, describe the steady-state error that would be produced for:
(a) A step input
(b) A ramp input
(c) A parabolic input
In each case, indicate the modifications in the system design which could be performed to reduce the steady-state error.

7.6. Given a system with a feedforward transfer function

$$G(s) = \frac{K(2s+1)}{s(s+1)(s^2 + 8s + 7)}$$

and a feedback transfer function

$$H(s) = 3s + 2$$

where K is a constant to be determined, describe the steady-state error that would be produced for:
(a) A step input
(b) A ramp input
(c) A parabolic input
In each case, indicate the modifications in the system design that could be performed to reduce the steady-state error.

7.7. A system with unity feedback has the following feedforward transfer function:

$$G(s) = \frac{K(3s + 1)}{s(s^2 + 10s + 9)(s + 1)^2}$$

(a) Identify the system type.

(b) Given a ramp input $R(t) = 5t$, determine the value of K for which the steady-state error will not exceed a value equal to 0.01.

(c) Determine the equivalent open-loop transfer function $T(s)$ for this system.

(d) For the input function $R(t)$ and the value of K found in part (b), determine the system response $C(t)$.

7.8. A system has the following feedforward transfer functions:

$$G_1(s) = \frac{K}{s} \qquad G_2(s) = \frac{2s + 1}{s(s + 3)}$$

with a feedback transfer function given by

$$H(s) = 3s + 5$$

(a) Identify the system type.

(b) Given a parabolic input $R(t) = 7t^2$, determine the value of K for which the steady-state error will not exceed a value equal to 0.01.

(c) Determine the equivalent open-loop transfer function $T(s)$ for this system.

(d) If $G_1(s)$ represents the transfer function for the controller, what type of control action is used in this system?

(e) For the input function $R(t) = 7t^2$ and the value of K found in part (b), determine the system response $C(t)$.

7.9. Describe two examples of actual systems in which proportional control is used. Use block diagram notation to represent these systems.

7.10. Describe two examples of actual systems in which integral control action is used. Use block diagram notation to represent these systems.

7.11. Describe two examples of actual systems in which proportional-plus-derivative control action is used. Use block diagram notation to represent these systems.

7.12. Describe two examples of actual systems in which proportional-plus-integral control action is used. Use block diagram notation to represent these systems.

7.13. Describe two examples of actual systems in which proportional-plus-derivative-plus-integral control action is used. Use block diagram notation to represent these systems.

7.14. Consider the following system represented in block diagram form:

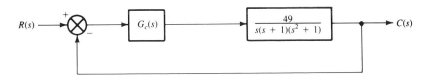

(a) Determine the equivalent open-loop transfer function $T(s)$ for this system in terms of the general controller transfer function $G_c(s)$.

(b) For $G_c = 3/s$, determine (1) the response function $C(t)$ and (2) the steady-state error, given a parabolic input $R(t) = 15t^2$.

(c) For $G_c = 3$, determine (1) the response function $C(t)$ and (2) the steady-state error, given a parabolic input $R(t) = 15t^2$.

(d) For $G_c = 3 + 2s$, determine (1) the response function $C(t)$ and (2) the steady-state error, given a parabolic input $R(t) = 15t^2$.

7.15. Consider the following system represented in block diagram form:

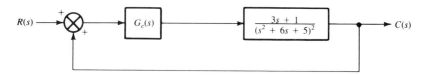

(a) Determine the equivalent open-loop transfer function $T(s)$ for this system in terms of the general controller transfer function $G_c(s)$.

(b) Identify the type of forcing function $R(s)$ for which the steady-state error will vanish if integral control is used in this system.

(c) For $G_c = 6/s$, determine (1) the response function $C(t)$ and (2) the steady-state error, given a parabolic input $R(t) = 3t^2$.

(d) For $G_c = 6/s$, determine (1) the response function $C(t)$ and (2) the steady-state error, given a ramp input $R(t) = 3t$.

(e) For $G_c = 6/s$, determine (1) the response function $C(t)$ and (2) the steady-state error, given a step input $R(t) = 3$.

7.16. For the system shown in Exercise 7.14, apply proportional-plus-integral control with

$$G_c(s) = 4 + \frac{3}{s}$$

Determine (1) the response function $C(t)$ and (2) the steady-state error for the cases in which one applies a forcing function $R(t)$ such that:

(a) $R(t) = 15$

(b) $R(t) = 15t$

(c) $R(t) = 15t^2$

7.17. For the system shown in Exercise 7.14, apply proportional-plus-derivative control with

$$G_c(s) = 4 + 3s$$

Determine (1) the response function $C(t)$ and (2) the steady-state error for the cases in which one applies a forcing function $R(t)$ such that:

(a) $R(t) = 15$

(b) $R(t) = 15t$

(c) $R(t) = 15t^2$

7.18. For the system shown in Exercise 7.14, apply proportional-plus-derivative-plus-integral control with

$$G_c(s) = 4 + 3s + \frac{3}{s}$$

Determine (1) the response function $C(t)$ and (2) the steady-state error for the cases in which one applies a forcing function $R(t)$ such that:
(a) $R(t) = 15$
(b) $R(t) = 15t$
(c) $R(t) = 15t^2$

7.19. Determine the steady-state error for a system with unity feedback and the feedforward transfer function $G(s)$ given in (a) or (b) below. The steady-state error should be determined for (1) a step input $R(t) = A$, (2) a ramp input $R(t) = At$, and (3) a parabolic input $R(t) = At^2$.

(a) $G(s) = \dfrac{5}{s(s+2)(s+4)}$

(b) $G(s) = \dfrac{1}{(s^2 + 8s + 7)(s^2 + 1)}$

7.20. Determine the response $C(t)$ and the steady-state error for the system described by the following block diagram:

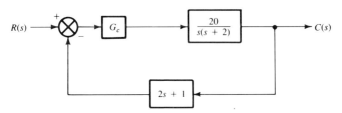

$R(s)$ is given by

$$R(s) = \frac{4}{s+2}$$

The system should be investigated for each of the following cases.
(a) $G_c = 3$ (proportional control)
(b) $G_c = 2/s$ (integral control)
(c) $G_c = 3 + 2/s$ (proportional-plus-integral control)
(d) $G_c = 3 + 4s$ (proportional-plus-derivative control)
(e) $G_c = 3 + 4s + 2/s$ (proportional-plus-derivative-plus-integral control)

7.21. Determine the response $C(t)$ and the steady-state error for the system described by the following block diagram:

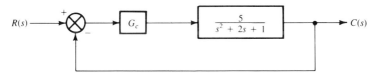

$R(s)$ is given by

$$R(s) = \frac{50}{s}$$

and

$$G_c(s) = 3 + K_d s + \frac{K_i}{s}$$

The system should be investigated for each of the following cases.
(a) $K_d = 0$, $K_i = 0$ (proportional control)
(b) $K_d = 1$, $K_i = 0$ (proportional-plus-derivative control)
(c) $K_d = 0$, $K_i = 2$ (proportional-plus-integral control)
(d) $K_d = 1$, $K_i = 2$ (proportional-plus-derivative-plus-integral control)

7.22. Evaluate the system described in Exercise 7.21 in accordance with the directions given therein; however, use the following input:

$$R(s) = \frac{50}{s + 2}$$

7.23. Evaluate the system described in Exercise 7.20 in accordance with the directions given therein; however, use the following input:

$$R(s) = \frac{4}{s}$$

7.24. Determine the transfer function $G_c(s)$ for the following system so that the steady-state error is equal to zero.

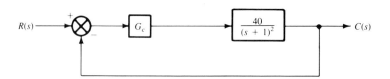

7.25. Determine the response $C(s)$ for the following system for each of the cases described below.

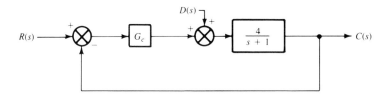

The input $R(s)$ equals $4/s$; the transfer function $G_c(s)$ is given by K/s.
(a) For the case in which there is no disturbance $D(s)$ acting on the system [i.e., $D(s) = 0$], evaluate $C(s)$ for the following values of K: $K = 0.1$, $K = 0.5$, $K = 1.0$, and $K = 5.0$.
(b) For the case in which $D(s) = 0.1/s$, evaluate $C(s)$ for each of the values of K given in part (a).
(c) For the case in which $D(s) = 1/s$, evaluate $C(s)$ for each of the values of K given in part (a).

(d) For the case in which $D(s) = 1/(s + 1)$, evaluate $C(s)$ for each of the values of K given in part (a).

7.26. Evaluate the system described in Exercise 7.25 in accordance with the directions given therein; however, use the following functions:

$$R(s) = \frac{4}{s^2 + 16} \qquad G_c(s) = \frac{K}{s}$$

7.27. Evaluate the system described in Exercise 7.25 in accordance with the directions given therein; however, use the following functions:

$$R(s) = \frac{4}{s} \qquad G_c(s) = K$$

7.28. Evaluate the system described in Exercise 7.25 in accordance with the directions given therein; however, use the following functions:

$$R(s) = \frac{4}{s^2 + 16} \qquad G_c(s) = K$$

7.29. Evaluate the system described in Exercise 7.25 in accordance with the directions given therein; however, use the following functions:

$$R(s) = \frac{4}{s} \qquad G_c(s) = Ks + 2$$

7.30. Evaluate the system described in Exercise 7.25 in accordance with the directions given therein; however, use the following functions:

$$R(s) = \frac{4}{s^2 + 16} \qquad G_c(s) = Ks + 2$$

7.31. Evaluate the system described in Exercise 7.25 in accordance with the directions given therein; however, use the following functions:

$$R(s) = \frac{4}{s} \qquad G_c(s) = Ks + 2 + \frac{1}{s}$$

7.32. Evaluate the system described in Exercise 7.25 in accordance with the directions given therein; however, use the following functions:

$$R(s) = \frac{4}{s^2 + 16} \qquad G_c(s) = Ks + 2 + \frac{1}{s}$$

Exercises 7.33 through 7.45 refer to the following system:

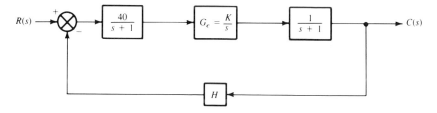

Assume that

$$R(s) = \frac{100}{s^2}$$

Determine the response $C(t)$ and the steady-state error in each of the following cases. Comment on the effect of K on $C(t)$ and the steady-state error.

7.33. Given that $H(s) = 1$.

7.34. Given that $H(s) = 0.2s + 1$.

7.35. Given that $H(s) = 2s + 1$.

For Exercises 7.36 through 7.38, use

$$G_c = \frac{K}{s} + 1$$

7.36. Given that $H(s) = 1$.

7.37. Given that $H(s) = 0.2s + 1$.

7.38. Given that $H(s) = 2s + 1$.

For Exercises 7.39 through 7.41, use

$$G_c = \frac{2}{s} + 1 + Ks$$

7.39. Given that $H(s) = 1$.

7.40. Given that $H(s) = 0.2s + 1$.

7.41. Given that $H(s) = 2s + 1$.

For Exercises 7.42 through 7.44, use

$$R(s) = \frac{100}{s}$$

7.42. Given that $G_c(s) = K/s$, $H(s) = 1$.

7.43. Given that $G_c(s) = (K/s) + 1$, $H(s) = 1$.

7.44. Given that $G_c(s) = Ks + 1 + (2/s)$, $H(s) = 1$.

7.45. Determine the steady-state error, in terms of the system parameters, for the speed control system described in Exercise 6.42 given:
   (a) A step input $R(s) = A/s$
   (b) A ramp input $R(s) = A/s^2$

7.46. Determine the steady-state error, in terms of the system parameters, for the speed control system described in Exercise 6.45 given:
   (a) A step input $R(s) = A/s$ and a disturbance $D(s) = B/s$
   (b) A ramp input $R(s) = A/s^2$ and a disturbance $D(s) = B/s$

8

STABILITY

What one man can invent another can discover.

Sherlock Holmes, The Adventure of the Dancing Men
by Sir Arthur Conan Doyle

8.1. OBJECTIVES

Upon completion of this chapter, the reader should be able to:

- Apply both the Routh–Hurwitz technique and the root-locus method in the analysis of system stability.
- Determine the number of roots (if any) of the characteristic equation for a system which lie to the right of the imaginary axis in the complex s-plane; such roots have positive real parts which lead to unbounded system responses (i.e., the system will not be stable).
- Evaluate the effect of a variable parameter upon system stability through the development of a root-loci diagram.
- Specify the value of a variable parameter which will produce marginal stability (a condition in which the system response is oscillatory with constant amplitude).

8.2 ROUTH–HURWITZ STABILITY CRITERION

Although we have several numerical techniques for determining the roots of the characteristic equation of a system (see Chapter 4), it would be convenient if we could quickly determine if a system is stable without developing a complete analysis of the system response $C(t)$. To some extent, the final-value theorem allows us to achieve this goal once we have obtained the Laplace transform $C(s)$ of the system response.

Recall that if a root of the characteristic equation is a real number a, the transient system response will contain a term that will behave exponentially in accordance with a multiplication factor e^{at}. Similarly, if two roots of the characteristic equation form a complex conjugate pair with values $(a \pm jb)$, the transient system response will include a term $Ce^{at} \sin(bt + \phi)$. If the real portion a of any of these roots is positive, the transient response will increase exponentially with time until the physical system collapses; that is, the system response will *not* remain finite for a bounded forcing function. We then state that such a system is *unstable*.

Routh and Hurwitz independently developed similar methods for determining the number of roots of the characteristic equation which have positive real parts or, equivalently, the number of roots that lie to the right of the imaginary axis in the complex s-plane.

The Hurwitz determinants (see Schwarzenbach and Gill, 1978) and the Routh criterion (Routh, 1877) provide the basis for quickly determining if a specific system design can be expected to produce stable behavior. The method allows us to determine the range of variation for a particular parameter (e.g., a gain K) that will produce a stable response; we may then, for example, minimize the steady-state error e_{ss} while maintaining stability.

The Routh–Hurwitz method is as follows: Given a system's characteristic equation in the s-domain (with zero initial conditions)

$$a_n s^n + a_{n-1} s^{n-1} + \cdots + a_1 s + a_0 = 0 \qquad (8.1)$$

a *necessary* condition for no roots of this expression to lie in the positive real half of the complex s-plane is that all coefficients $(a_n, a_{n-1}, \ldots, a_1, a_0)$ have the same sign (positive or negative) and that all coefficients have nonzero values. [If a_n is negative, we will assume that equation (8.1) is multiplied by a factor equal to -1, so that we may always treat a_n as a positive number. The necessary condition for stability then states that all coefficients a_i must be positive and nonzero.] For example, given the characteristic equations

$$4s^4 + 3s^3 + 2s + 1 = 0 \qquad (8.2a)$$

and

$$2s^3 - s^2 + s + 5 = 0 \qquad (8.2b)$$

the necessary condition for stability indicates that both of the expressions represent systems in which stability will not be achieved. The first equation (8.2a) is "missing" a term in s^2 (i.e., a_2 equals zero), whereas the second expression (8.2b) has coefficients which vary in sign $(+2, -1, +1, +5)$. An equation of the form

$$5s^4 + 3s^3 + 2s^2 + 5s + 2 = 0 \qquad (8.2c)$$

may represent the behavior of a stable system; a sufficient condition for stability is needed if we are to determine if the corresponding system *is* stable. A Routh–Hurwitz array or table will provide us with the necessary information to formulate a sufficient condition for stability.

We construct the Routh–Hurwitz table for the nth-order characteristic equation (8.1), with $(n + 1)$ rows, as follows:

Row 1 = s^n	a_n	a_{n-2}	a_{n-4}	$\cdots$
Row 2 = s^{n-1}	a_{n-1}	a_{n-3}	a_{n-5}	$\cdots$
Row 3 = s^{n-2}	b_1	b_2	b_3	$\cdots$
Row 4 = s^{n-3}	c_1	c_2	c_3	$\cdots$
Row 5 = s^{n-4}	d_1	d_2	d_3	$\cdots$
$\vdots$	$\vdots$	$\vdots$	$\vdots$	
Row n = s^1	$\cdot$	$\cdot$	$\cdot$	
Row $(n + 1)$ = s^0	$\cdot$	$\cdot$	$\cdot$	

The elements of the first two rows of the Routh–Hurwitz array consist of the coefficients of the characteristic equation, where a_n, a_{n-2}, ... form the first row and where a_{n-1}, a_{n-3}, ... are the elements of the second row. The third-row elements are constructed in accordance with the relations

$$b_1 = \frac{(a_{n-1}a_{n-2}) - (a_n a_{n-3})}{a_{n-1}} \tag{8.3a}$$

$$b_2 = \frac{(a_{n-1}a_{n-4}) - (a_n a_{n-5})}{a_{n-1}} \tag{8.3b}$$

$$b_3 = \frac{(a_{n-3}a_{n-6}) - (a_{n-2}a_{n-7})}{a_{n-1}} \tag{8.3c}$$

$$\vdots \qquad\qquad \vdots \qquad\qquad \vdots$$

where we divide by the value a_{n-1} in order to control the size of these third-row elements b_i. Each b_i is the result of a cross-multiplication of certain elements in the preceding two rows. Except in the construction of b_1, elements are taken from the columns adjacent to that in which b_i is located; for example, for the construction of b_2, elements are taken from column 1 (a_n and a_{n-1}) and column 3 (a_{n-4} and a_{n-5}) and then cross-multiplied:

$$b_2 = \frac{1}{a_{n-1}} \begin{vmatrix} a_n & a_{n-4} \\ a_{n-1} & a_{n-5} \end{vmatrix}$$

$$= \frac{(a_{n-1}a_{n-4}) - (a_n a_{n-5})}{a_{n-1}}$$

Note that cross-multiplication from the lower-left corner to the upper-right corner results in a *positive* term (exactly the opposite of cross-multiplication of elements in a standard determinant).

The elements (c_i) in the fourth row of the array are constructed in a similar fashion, in accordance with the relations

$$c_1 = \frac{(b_1 a_{n-3}) - (a_{n-1} b_2)}{b_1} \tag{8.4a}$$

$$c_2 = \frac{(b_1 a_{n-5}) - (a_{n-1} b_3)}{b_1} \tag{8.4b}$$

$$c_3 = \frac{(b_2 a_{n-7}) - (a_{n-3} b_4)}{b_1} \tag{8.4c}$$

$$\vdots \qquad \vdots \qquad \qquad \vdots$$

For the fifth-row elements d_i:

$$d_1 = \frac{(c_1 b_2) - (b_1 c_2)}{c_1} \tag{8.5a}$$

$$d_2 = \frac{(c_1 b_3) - (b_1 c_3)}{c_1} \tag{8.5b}$$

$$d_3 = \frac{(c_2 b_4) - (b_2 c_4)}{c_1} \tag{8.5c}$$

$$\vdots \qquad \vdots \qquad \qquad \vdots$$

Each row is constructed until zero-value elements are repeatedly obtained.

The Routh criterion for *all* roots to have negative real parts (i.e., lie to the left of the imaginary axis in the complex s-plane) is that *all elements in the first column of the table must be nonzero and have the same sign*, that is, if $a_n > 0$, then a_{n-1}, b_1, c_1, ... must all be positive and nonzero. (Of course, since each of these elements depend on elements in the previous rows, all rows of the array must be constructed.)

Furthermore, *the number of sign changes (if any) among the elements in the first column of the array is identical to the number of roots in the positive real half of the complex s-plane.*

Example 8.1

Given the following characteristic equation for a system

$$5s^4 + 3s^3 + 2s^2 + 5s + 2 = 0 \tag{8.6}$$

we construct the corresponding Routh–Hurwitz table:

s^4	5	2	2
s^3	3	5	
s^2	$-\frac{19}{3}$	2	
s^1	$\frac{113}{19}$	0	
s^0	2		

We note that there are two sign changes in column 1:

$$+3 \rightarrow -\tfrac{19}{3}$$

$$-\tfrac{19}{3} \rightarrow +\tfrac{113}{19}$$

Therefore, we conclude that there are two roots of this polynomial expression in s which lie in the right half of the complex s-plane; the corresponding system is unstable.

We now wish to consider the following special cases:

1. If a zero appears as the first element in a row in which at least one other element has a nonzero value, the terms in the next row cannot be computed directly from the above Routh–Hurwitz algorithm since we would be dividing by the zero-value element. For example, the expression

$$s^4 + 2s^3 + s^2 + 2s + 2 = 0 \tag{8.7}$$

produces a table of the form

s^4	1	1	2
s^3	2	2	
s^2	0	2	
s^1	$-\infty$	$\rightarrow$	$\rightarrow$ *cannot continue!*

In such a case, one may (1) replace the zero element in the first column with a small number ϵ in order to continue the construction of the array (see Raven, 1978) or (2) replace the variable s in the characteristic equation by a new variable σ^{-1}, which then produces a new (transformed) equation in which the order of the coefficients in the original expression has been reversed (see Schwarzenbach and Gill, 1978). The Routh–Hurwitz table is then constructed for this new equation; if no roots of this transformed equation in σ are found to lie in the positive real half of the complex σ-plane, then no roots of the original equation lie in the positive real half of the complex s-plane. Continuing the example, we perform the transformation

$$s \rightarrow \frac{1}{\sigma}$$

which then transforms the expression

$$s^4 + 2s^3 + s^2 + 2s + 2 = 0$$

into

$$\left(\frac{1}{\sigma}\right)^4 + 2\left(\frac{1}{\sigma}\right)^3 + \left(\frac{1}{\sigma}\right)^2 + 2\left(\frac{1}{\sigma}\right) + 2 = 0 \tag{8.8a}$$

or, upon multiplication by σ^4 and rearrangement of terms:

$$2\sigma^4 + 2\sigma^3 + \sigma^2 + 2\sigma + 1 = 0 \tag{8.8b}$$

The corresponding Routh–Hurwitz array is

$$
\begin{array}{c|ccc}
\sigma^4 & 2 & 1 & 1 \\
\sigma^3 & 2 & 2 & \\
\sigma^2 & -1 & 1 & \\
\sigma^1 & 4 & 0 & \\
\sigma^0 & 1 & &
\end{array}
$$

Two sign changes are noted in column 1 of this table; two roots of the original characteristic equation (8.7) can be expected to lie on the right half of the complex s-plane.

2. If a row is composed of all zero-value elements, one or more pairs of roots are symmetrically located about the origin of the s-plane. Of particular interest is the case in which row n (corresponding to s^1) of the array is composed entirely of zero-value elements with nonzero elements in row ($n - 1$); there is then a pair of complex conjugate roots located along the imaginary axis of the complex s-plane which indicate that the transient system response will oscillate with constant amplitude—that is, a case of "marginal stability." To determine these symmetrically located roots, one simply constructs an *auxiliary equation* with coefficients equal to the elements appearing in the row above the "zero-elements-only" row (the order of this auxiliary equation is identical to the power of s corresponding to the row of coefficients for this equation). The roots of the auxiliary equation *are* the symmetrically located roots of the characteristic equation that we seek. Furthermore, differentiation of the auxiliary equation with respect to s will produce coefficients that can be used in place of the zero-value elements in the Routh–Hurwitz table; construction of the table can then be continued.

Example 8.2

Consider the characteristic equation

$$5s^3 + 2s^2 + 15s + 6 = 0 \tag{8.9}$$

for which we obtain the following Routh–Hurwitz array:

$$
\begin{array}{c|cc}
s^3 & 5 & 15 \\
s^2 & 2 & 6 \\
s^1 & 0 & 0 \\
s^0 & \rightarrow & \rightarrow \; cannot \; continue!
\end{array}
$$

The auxiliary equation is obtained from the s^2 row:

$$2s^2 + 6 = 0 \tag{8.10}$$

The roots of this equation are

$$r_1, r_2 = \pm j(3)^{1/2}$$

Furthermore, differentiation of this auxiliary equation gives

$$4s + 0 = 0 \tag{8.11}$$

so that the Routh–Hurwitz array can be completed:

$$
\begin{array}{c|cc}
s^3 & 5 & 15 \\
s^2 & 2 & 6 \\
s^1 & 4 & 0 \\
s^0 & 6 &
\end{array}
$$

No sign changes are found among the nonzero elements of the first column of this table; therefore, no roots of the characteristic equation lie in the right half of the complex s-plane.

We may also factor the original equation by the auxiliary equation:

$$5s^3 + 2s^2 + 15s + 6 = (2s^2 + 6)(2.5s + 1)$$

indicating that the roots of the characteristic equation are

$$r_1 = +j(3)^{1/2} \tag{8.12a}$$

$$r_2 = -j(3)^{1/2} \tag{8.12b}$$

$$r_3 = -0.4 \tag{8.12c}$$

The system is marginally stable, with oscillations of constant amplitude with a natural frequency equal to $(3)^{1/2}$ radians per second.

3. Finally, one may wish to determine if any roots of the characteristic equation lie in the left (negative real) half of the complex s-plane but within a horizontal distance κ of the imaginary axis. (Since those roots nearest the imaginary axis have the smallest real values, they produce terms in the transient system response which will exponentially damp most slowly to zero. Such terms will then dominate a stable transient response.)

A simple substitution of $(s + \kappa)$ for s in the characteristic equation will then produce an expression which, through application of the Routh–Hurwitz criterion, can be analyzed to determine the number of roots located to the right of a vertical line intersecting the horizontal real axis at $(-\kappa)$ in the complex s-plane (see Raven, 1978).

A principal application of the Routh–Hurwitz method is to determine a limiting value of a variable parameter for which the system response remains stable.

Example 8.3

Consider the transfer function $T_{\mathrm{PDI}}(s)$ of the system shown in Figure 7.13, in which proportional-plus-derivative-plus-integral control is used:

$$T_{\mathrm{PDI}}(s) = \frac{KK_d s^2 + KK_p s + KK_i}{s^3 + (KK_d + p)s^2 + (KK_p + q)s + KK_i} \tag{7.68}$$

• •

The characteristic equation for this system is then

$$s^3 + (KK_d + p)s^2 + (KK_p + q)s + KK_i = 0 \qquad (8.13)$$

Let us assume that all system parameters except the integral gain K_i are fixed, with values given by

$$K_p = 5$$
$$K_d = 1$$
$$K = 3$$
$$p = 7$$
$$q = 10$$

The characteristic equation is then

$$s^3 + 10s^2 + 25s + 3K_i = 0 \qquad (8.14)$$

We will determine the maximum value of K_i for which the system's transient response remains stable (i.e., finite).

A necessary condition for stability is that all coefficients in the expression (8.14) be nonzero and have the same sign; hence K_i must be greater than zero.

The Routh–Hurwitz array is then constructed:

$$
\begin{array}{c|cc}
s^3 & 1 & 25 \\
s^2 & 10 & 3K_i \\
s^1 & \dfrac{250 - 3K_i}{10} & 0 \\
s^0 & 3K_i &
\end{array}
$$

Thus the element $(250 - 3K_i)/10$ must be greater than zero for a stable response or

$$K_i < \tfrac{250}{3}$$

If $K_i = 250/3$, the system is marginally stable with two roots of the characteristic equation lying on the imaginary axis. The auxiliary equation is obtained from the table for this special case where $(250 - 3K_i)$ vanishes:

$$10s^2 + 3K_i = 10s^2 + 250$$
$$= 0 \qquad (8.15)$$

so that the two roots on the imaginary axis are

$$r_1, r_2 = \pm j(5)$$

Example 8.4

The relationship between a predator and its prey can be approximately described by the following equations (Dorf, 1974):

$$\frac{dx_1}{dt} = ax_1 - \alpha x_2 = Dx_1$$

$$\frac{dx_2}{dt} = bx_1 - \beta x_2 = Dx_2$$

where $x_1 \equiv$ population of the prey

$\quad x_2 \equiv$ population of predators

$\quad D \equiv d/dt$

and where a, b, α, and β are constants for the system. The first equation states that the rate dx_1/dt of change in the prey's population is directly proportional to x_1; this rate of change decreases as the population of the predator increases (in accordance with the term $-\alpha x_2$). Similarly, the second equation describes the relationship between the rate dx_2/dt of change in the predator population, x_1 and x_2; the rate increases with an increasing population x_1 of the prey and decreases with an increasing population x_2 of (competitive) predators. If we multiply the first equation by the quantity $(D + \beta)$, the second equation by $(-\alpha)$, and add the results, we obtain a single second-order equation in x_1:

$$[D^2 + (\beta - a)D + (\alpha b - a\beta)]x_1 = 0$$

(i.e., we have applied the method of multiplication by common coefficients described in Section 2.6). We may then obtain the characteristic equation via a Laplace transformation:

$$s^2 + (\beta - a)s + (\alpha b - a\beta) = 0$$

The corresponding Routh–Hurwitz array is then constructed:

$$
\begin{array}{c|cc}
s^2 & 1 & \alpha b - a\beta \\
s^1 & \beta - a & \\
s^0 & \alpha b - a\beta &
\end{array}
$$

The necessary and sufficient conditions for stability state that the coefficients in the second-order equation must be nonzero and that

$$\beta - a > 0$$

$$\alpha b - a\beta > 0$$

The second inequality is of particular interest. It states that the product of the coefficients (a and β) to increase $\dot{x}_1$ and decrease $\dot{x}_2$ must be less than the product of the coefficients (α and b) to decrease $\dot{x}_1$ and increase $\dot{x}_2$ if the population x_1 of the prey is to be bounded.

8.3 ROOT-LOCUS METHOD

The Routh–Hurwitz technique allows us to determine limiting values of a variable parameter which maintain stability within a system; however, this technique does not provide us with the specific values of the roots of the system's characteristic equation. The numerical techniques of Chapter 4 (and other numerical techniques which we have not discussed) do allow us to determine the roots of a system's characteristic equation with the aid of a computer.

During the latter part of the nineteenth century, graphical methods for analyzing descriptive system equations were developed (Tolle, 1895, 1896; Mayr, 1970); finally, W. R. Evans (1948, 1950, 1954) generated his root-locus method, which allows one to determine graphically the root values of the characteristic equation for the case in which a particular system parameter is undergoing variation, thereby causing each root value to vary also.

Evans's method is used to plot quickly the loci of root values of the characteristic equation for the range of variation in the parameter that is to be manipulated. Such plotting can be done manually with relative ease and without the aid of a computer. In addition, the variable parameter is frequently a control gain (e.g., K_p, K_i); the root-locus method allows one to choose a value for this parameter which ensures stable operation of the system together with other desired objectives (e.g., minimum steady-state error e_{ss}, minimum settling time, minimum overshoot).

8.3.1 Root-Loci Plots

As an introduction to the concept of root loci [based on a discussion by Schwarzenbach and Gill (1978)], consider the system with proportional control shown in Figure 8.1. The characteristic equation for this system is

$$1 + K_p G(s) = 1 + \frac{K_p K_1}{s^2 + ps + q}$$
$$= 0 \tag{8.16}$$

or, equivalently,

$$s^2 + ps + q + K_p K_1 = 0 \tag{8.17}$$

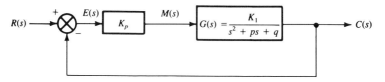

Figure 8.1   Example system with proportional control.

For simplicity, we introduce a new parameter, K', such that

$$K' \equiv q + K_p K_1 \tag{8.18}$$

and we choose p equal to unity ($p = 1$) so that the characteristic equation (8.17) becomes

$$s^2 + s + K' = 0 \tag{8.19}$$

We seek the loci of roots for this equation in the complex s-plane for all possible values of K' (realistically, we wish to vary K' by varying K_p while keeping q constant).

We begin with K' set equal to zero, thereby giving

$$s^2 + s = 0 \tag{8.20}$$

with corresponding roots

$$r_1, r_2 = 0, -1 \tag{8.21}$$

We next construct a table (Table 8.1) of root values for various values of K':

TABLE 8.1   Root Values (for Specific K' Values) of Characteristic Equation (8.19)

K'	r_1	r_2
0.00	0.000	1.00
0.25	-0.500	-0.500
1.00	$-0.5 + j\,(0.866)$	$-0.5 - j\,(0.866)$
2.00	$-0.5 + j\,(1.323)$	$-0.5 - j\,(1.323)$
6.00	$-0.5 + j\,(2.400)$	$-0.5 - j\,(2.400)$
20.00	$-0.5 + j\,(4.444)$	$-0.5 - j\,(4.444)$
50.00	$-0.5 + j\,(7.053)$	$-0.5 - j\,(7.053)$
75.00	$-0.5 + j\,(8.646)$	$-0.5 - j\,(8.646)$
100.0	$-0.5 + j\,(9.987)$	$-0.5 - j\,(9.987)$
$\vdots$	$\vdots$	$\vdots$

Figure 8.2 presents the loci of roots in the complex s-plane for this system which corresponds to such variations in K'. Notice that K' can be increased without limit (theoretically) while the system behavior remains stable (i.e., no positive real root values appear as K' is varied). If K' is set equal to zero, one root lies at the origin of the complex s-plane so that the system's transient response will contain a constant term. For K' values less than or equal to 0.25, the roots lie along the (negative) real axis, so that the system response will not include oscillations (i.e., the system is overdamped). Critical damping corresponds to a K' value of 0.25; underdamped behavior is obtained for K' values greater than 0.25 (oscillatory behavior is obtained).

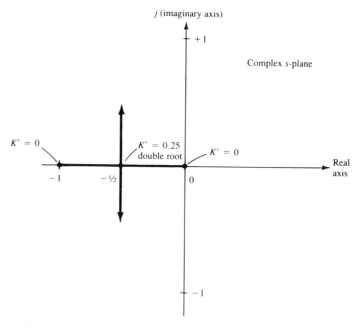

Figure 8.2 Complex s-plane and root loci for example system. (Based on a similar example by Schwarzenbach and Gill, 1978.)

This example demonstrates how we may generate a graphical description of a system's root loci in the complex s-plane which corresponds to variations in a given system parameter. We now wish to describe Evans's method for quickly generating such graphical descriptions of root loci.

8.3.2 Root-Loci Procedure

Recall that the characteristic equation for a system with closed-loop feedback (Figure 8.3) is as follows:

$$1 + G_1 G_2 \cdots G_j H_1 H_2 \cdots H_l = 0 \qquad (8.22)$$

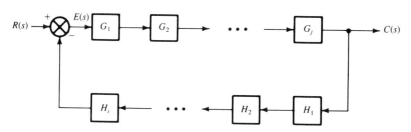

Figure 8.3 System with closed-loop feedback.

If we write the quantity

$$G_1 G_2 \cdots G_j H_1 H_2 \cdots H_l = K \frac{N}{D} \qquad (8.23)$$

where K is a constant, N represents the polynomial of order m in s which forms the numerator of this quantity, and D is the polynomial of order n in s which forms the denominator of the product of the feedforward transfer functions $G_1, G_2, \ldots, G_j$ and the feedback transfer functions $H_1, H_2, \ldots, H_l$. We may then write this quantity in terms of the individual factors in s:

$$K \frac{N}{D} = K \frac{(s - z_1)(s - z_2) \cdots (s - z_m)}{(s - p_1)(s - p_2) \cdots (s - p_n)} \qquad (8.24)$$

The roots $z_1, z_2, \ldots, z_m$ of the polynomial in the numerator of this expression are the *zeros* of KN/D (i.e., the values of s for which KN/D vanishes). The roots $p_1, p_2, \ldots, p_l$ of the polynomial forming the denominator D of this expression are the *poles* (i.e., the values of s for which KN/D becomes infinite). [For all physical systems, $n \geq m$ (see Dorf, 1974).]

The characteristic equation can then be expressed as

$$-1 = K \frac{(s - z_1) \cdots (s - z_m)}{(s - p_1) \cdots (s - p_n)} \qquad (8.25)$$

from which we obtain the *magnitude criterion*:

$$1 = K \frac{|s - z_1||s - z_2| \cdots |s - z_m|}{|s - p_1||s - p_2| \cdots |s - p_n|} \qquad (8.26)$$

and the *angle criterion* (in units of radians):

$$\{\angle(s - z_1) + \angle(s - z_2) + \cdots + \angle(s - z_m)\}$$
$$- \{\angle(s - p_1) + \angle(s - p_2) + \cdots + \angle(s - p_n)\} = (2i + 1)\pi$$

where $\angle(s - z_j)$ represents the angle between a point s and the zero z_j in the complex s-plane and where i is an integer [see Dorf (1974) for details]. The magnitude criterion allows one to determine the constant K which corresponds to a particular point s on the locus of roots; the angle criterion allows one to generate the trace of the locus.

Adapting the discussion by Dorf (1974) of the procedure for the root-locus method, we may identify the following rules, which are used to plot the root loci:

Rule 1.   The number of root loci equals the number of (open-loop) poles n.

Rule 2.   A root locus begins at each pole p_i and terminates at either one of the zeros z_j or at infinity. The number of loci terminating at infinity equals $(n - m)$ (i.e., the difference between the number of poles and the number of zeros). A double pole (a pole value that corresponds to two roots) will have two loci emerging from it; a pole of order q will have q loci emerging from it. Similarly, a zero of order r will have r loci terminating at its location. (One may think of poles as "*sources*" of root loci and zeros as "*sinks*" for loci.)

Rule 3. One may determine which portions of the real axis in the complex s-plane contain part of the root loci by applying the following rule: If the total number of zeros and poles located on the real axis to the *right* of a given location is odd, that location on the axis is part of a root locus. (Remember that poles and zeros of multiple order must be counted as multiple poles or zeros.)

Rule 4. A total of $(n - m)$ loci asymptotically approach $(n - m)$ straight lines in the s-plane (where these loci terminate at infinity), each of which is directed toward a "center" γ of the poles and zeros given by

$$\gamma = \frac{\sum\limits_{j=1}^{n} p_j - \sum\limits_{i=1}^{m} z_i}{n - m} \tag{8.27}$$

where these straight lines are directed at angles (relative to the real axis) given by

$$\pi \frac{2k + 1}{n - m} \quad \text{radians}$$

where $k = 0, 1, 2, \ldots, n - m - 1$. These lines are spaced at angles equal to $[2\pi/(n - m)]$ radians relative to one another.

Rule 5. Two root loci, emerging from adjacent poles on the real axis, intersect and then depart from this axis (at angles of $\pm\pi/2$) at a so-called *breakaway point* s, which is given by

$$\sum\limits_{i=1}^{m} \frac{1}{s - z_i} = \sum\limits_{j=1}^{n} \frac{1}{s - p_j} \tag{8.28}$$

Rule 6. The q loci, emerging from a qth-order pole p_a, depart at angles θ given by

$$\theta = \frac{1}{q}\left[(2k + 1)\pi + \sum\limits_{i=1} \angle(p_a - z_i) - \sum\limits_{\substack{j=1 \\ j \neq a}} \angle(p_a - p_j)\right] \tag{8.29}$$

where $k = 0, 1, \ldots, q - 1$.

Example 8.5

As an example of the application of the root-locus method, consider the system with integral control action shown in Figure 8.4. The characteristic equation for this system is

$$1 + \frac{KK_i}{s(s + 2)(s + 5)} = 0 \qquad (8.30a)$$

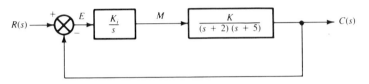

Figure 8.4   Example system.

or, if we introduce $K' \equiv KK_i$,

$$1 + \frac{K'}{s(s + 2)(s + 5)} = 0 \qquad (8.30b)$$

To construct the root loci for this system, we may apply the rules given earlier.

Rule 1:

$$K\frac{N}{D} = \frac{K'}{s(s + 2)(s + 5)} \qquad (8.31)$$

so that the three poles for the system are

$$p_1 = 0 \qquad (8.32a)$$
$$p_2 = -2 \qquad (8.32b)$$
$$p_3 = -5 \qquad (8.32c)$$

(There are no zeros z_i for this system.) Therefore, there are three loci which emerge from the three poles.

Rule 2: Since

$$n - m = 3 \qquad (8.33)$$

all three loci terminate at infinity.

Rule 3: The portions of the real axis that contain part of the loci are (a) between 0.0 and -2.0, and (b) between -5.0 and $-\infty$ (see Figure 8.5).

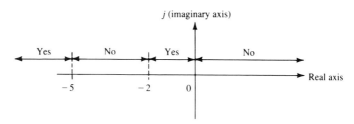

Figure 8.5   Loci on real axis.

Rule 4: The center γ is given by

$$\gamma = \frac{\sum\limits_{j=1}^{n} p_j - \sum\limits_{i=1}^{m} z_i}{n-m}$$

$$= \frac{(0 - 2 - 5) - (0)}{3 - 0} \tag{8.34}$$

$$= -\frac{7}{3}$$

The three loci asymptotically approach straight lines which are directed toward the point γ at angles given by

$$\pi\frac{2k+1}{n-m} = \frac{\pi}{3}, \pi, \frac{5\pi}{3} \text{ radians}$$
$$= 60°, 180°, 300° \tag{8.35}$$

where $k = 0$, 1, and 2 (see Figure 8.6). As a result, we expect that the focus emerging from the (leftmost) pole at -5 is directed toward negative infinity along the real axis (see Rule 3 and its results).

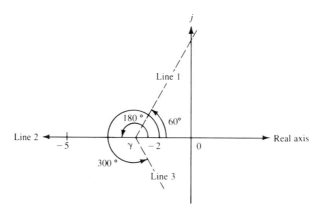

Figure 8.6   Asymptotically approached straight lines.

Rule 5: The breakaway point for the two adjacent loci at 0 and -2 is given by

$$\sum_{i=1}^{m} \frac{1}{s - z_i} = \sum_{j=1}^{n} \frac{1}{s - p_j}$$

or

$$0 = \frac{1}{s} + \frac{1}{s+2} + \frac{1}{s+5} \qquad (8.36)$$

from which we obtain

$$3s^2 + 14s + 10 = 0 \qquad (8.37)$$

Two solutions are possible for this quadratic expression:

$$\begin{aligned} s_1, s_2 &= -\frac{14}{6} \pm \frac{(76)^{1/2}}{6} \\ &= -2.333 \pm 1.453 \qquad (8.38) \\ &= -0.88, -3.786 \end{aligned}$$

The breakaway point *must* lie on a portion of the real axis which, in accordance with Rule 3 and its results, contains part of the root loci; therefore, the breakaway point must be located at -0.88. Figure 8.7 shows our knowledge of the loci at this point in our analysis.

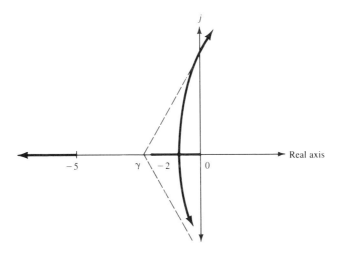

Figure 8.7   Rough sketch of root loci, developed from the application of rules 1 through 5.

Rule 6: The order of each pole is unity, so that

$$\theta = 0° \text{ or } 180°$$

as expected for the angle of emergence of each loci.

We may also use the Routh–Hurwitz criteria to determine the intersection of any loci with the imaginary axis (points of marginal stability).

The characteristic equation for this example system can be written in the form

$$s^3 + 7s^2 + 10s + K' = 0 \tag{8.39}$$

The Routh–Hurwitz table is then

$$
\begin{array}{c|cc}
s^3 & 1 & 10 \\
s^2 & 7 & K' \\
s^1 & 10 - \dfrac{K'}{7} & \\
s^0 & K' &
\end{array}
$$

Hence, for stable behavior,

$$10 - \frac{K'}{7} > 0$$

or

$$K' < 70$$

If

$$K' = 70$$

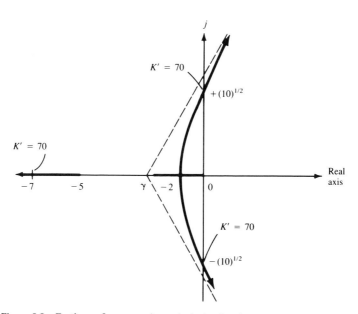

Figure 8.8   Further refinement of root-loci plot for the system of Example 8.5.

we have marginal stability with two roots located on the imaginary axis; the auxiliary equation

$$7s^2 + K' = 7s^2 + 70$$
$$= 0 \qquad (8.40)$$

provides us with the values of these two roots:

$$r_1, r_2 = \pm j(10)^{1/2}$$

with the corresponding third root of the characteristic equation given by

$$s^3 + 7s^2 + 10s + 70 = 7(s^2 + 10)\left(\frac{s}{7} + 1\right) \qquad (8.41)$$

or

$$r_3 = -7$$

(see Figure 8.8).

Finally, the magnitude and angle criterion may be used to determine specific root values or the loci which correspond to particular values of K' (and vice versa).

8.4 REVIEW

In summary, we have reviewed the following topics, facts, relationships, or concepts in this chapter.

- A system will be stable (i.e., it will produce a finite response to a given bounded forcing function) if all roots of its characteristic equation lie on the (negative real) left side of the imaginary axis in the complex s-plane. A *necessary* (but not sufficient) condition for stable behavior is that all coefficients of the system's characteristic equation be nonzero and have the same sign.

- The Routh–Hurwitz method provides us with a *sufficient* condition for stability: All elements in the first column of the Routh–Hurwitz array must be nonzero and must have the same sign if all roots of the characteristic equation are to have real parts which are negative.

    The number of sign changes among the elements of this first column equals the number of roots that lie on the (positive real) right side of the complex s-plane.

    Special cases which were considered include:

    (a) A zero appearing in the first column of a row (with at least one other element in that row being nonzero); in this case, we may replace the variable s by a new variable σ^{-1} which allows us to continue the construction of the Routh–Hurwitz array.

    (b) A row is composed of all zero-value elements; one or more pairs of roots are then symmetrically located about the origin of the s-plane. The *auxiliary equation*, obtained from the row preceding that with all zero-value elements, can be used to evaluate these pairs of symmetrically located roots. If the row of zero-value elements is the *n*th row (corresponding to s^1), the pair of symmetrically located roots lie along the imaginary axis in the s-plane and the system has *marginal stability*. Furthermore, differentiation of the auxiliary equation allows us to continue construction of the Routh–Hurwitz array.

(c) Substitution of s by $(s + \kappa)$ in the characteristic equation, followed by application of the Routh–Hurwitz method, allows one to determine the number of roots which lie to the left of the vertical line passing through the point at $-\kappa$ on the horizontal real axis in the complex s-plane.

A principal application of the Routh–Hurwitz method is to determine a limiting value of a variable parameter for which the system response remains finite.

• The root-locus method allows us to develop quickly a graphical description of the root loci in the complex s-plane for a range of values of a variable system parameter. The six rules of the method are used to identify the breakaway point(s), the center point γ for asymptotic straight lines, and so on.

EXERCISES

Exercises 8.1 through 8.10 refer to the following system; in each case, *sketch the root-locus diagram* for the system and describe the stability of the system behavior. Use the transfer function $G(s)$ given in the problem. Assume a positive value for K in all cases.

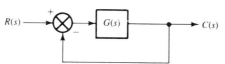

8.1. Given that

$$G(s) = \frac{K}{(s + 1)(s + 2)}$$

8.2. Given that

$$G(s) = \frac{K(2s + 1)}{s(s^2 + 2s + 1)}$$

8.3. Given that

$$G(s) = \frac{K}{s(s + 1)(s + 5)^2}$$

8.4. Given that

$$G(s) = \frac{K(4s + 1)(s + 2)}{s^3 + 2s^2 + s + 10}$$

8.5. Given that

$$G(s) = \frac{K}{s(s + 1)(s^2 + 5s + 2)}$$

8.6. Given that

$$G(s) = \frac{2K + 1}{(s^2 + 2s + 0.5)(s^2 + s + 0.1)}$$

8.7. Given that

$$G(s) = \frac{K(s + 1)(s + 5)}{(s^2 + s + 0.5)(s^3 + 2s^2 + s + 5)}$$

8.8. Given that

$$G(s) = \frac{10 + Ks}{(s + 2)^2(s^2 + 2s + 0.25)}$$

8.9. Given that

$$G(s) = \frac{2K}{(s + 1)(s^2 + 3s + 0.75)}$$

8.10. Given that

$$G(s) = \frac{2 + Ks}{10(s + 1)(s - 2)(s^2 + 3s + 3)}$$

Use the Routh-Hurwitz criterion to evaluate the stability of the systems for which the characteristic equations are given in Exercises 8.11 through 8.15.

8.11. $2s^2 + 4s - 1 = 0$

8.12. $8s^3 + 8s^2 + 4s + 2 = 0$

8.13. $16s^5 + 8s^4 + 2s^3 + s^2 + 0.25s + 10 = 0$

8.14. $s^4 + 3s^3 + 4s^2 + s + 8 = 0$

8.15. $s^4 + 2s^3 + 2s^2 + 2s + 2 = 0$

Use the Routh-Hurwitz criterion to evaluate the following characteristic equations. Determine the values (if any) of K for which the corresponding system behavior will be stable.

8.16. $s^3 + Ks^2 + s + 16 = 0$

8.17. $Ks^3 + 10s^2 + 8s + 10 = 0$

8.18. $Ks^5 + 8s^4 + 3s^3 + 8s^2 + 10s + 1 = 0$

8.19. $s^4 + (K + 6)s^3 + Ks^2 + 10s + 6 = 0$

8.20. $s^5 + 2s^4 + 3s^3 + (K + 10)s^2 + 10s + 3K = 0$

For the case of unity feedback, obtain the characteristic equation and apply the Routh–Hurwitz stability criterion to evaluate the system described in the exercise identified below.

8.21. Exercise 8.1   **8.22.** Exercise 8.2

8.23. Exercise 8.3   **8.24.** Exercise 8.4

8.25. Exercise 8.5   **8.26.** Exercise 8.6

8.27. Exercise 8.7   **8.28.** Exercise 8.8

8.29. Exericse 8.9   **8.30.** Exercise 8.10

For the case in which the feedback transfer function $H(s)$ is given by

$$H(s) = 2s + 1$$

obtain the characteristic equation and apply the Routh–Hurwitz stability criterion to evaluate the system described in the exercise identified below.

8.31. Exercise 8.1   **8.32.** Exercise 8.2

8.33. Exercise 8.3   **8.34.** Exercise 8.4

8.35. Exercise 8.5   **8.36.** Exercise 8.6

8.37. Exercise 8.7   **8.38.** Exercise 8.8

8.39. Exercise 8.9   **8.40.** Exercise 8.10

For Exercises 8.41 through 8.60, refer to the following closed-loop system:

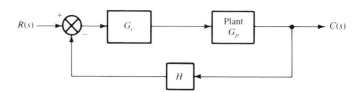

Sketch the root-locus diagram for each of the following cases; also, describe the stability of the system behavior. Assume that $K \geq 0$ in all cases.

8.41. Given that

$$G_c(s) = K \qquad G_p(s) = \frac{4}{s^2 + 5s + 2} \qquad H(s) = 1$$

8.42. Given that

$$G_c(s) = \frac{K}{s} \qquad G_p(s) = \frac{4}{s^2 + 5s + 2} \qquad H(s) = 1$$

8.43. Given that

$$G_c(s) = K(s + 1) \qquad G_p(s) = \frac{4}{s^2 + 5s + 2} \qquad H(s) = 1$$

8.44. Given that

$$G_c(s) = K\left(1 + \frac{1}{s}\right) \qquad G_p(s) = \frac{4}{s^2 + 5s + 2} \qquad H(s) = 1$$

8.45. Given that

$$G_c(s) = K\left(1 + \frac{1}{s} + s\right) \qquad G_p(s) = \frac{4}{s^2 + 5s + 2} \qquad H(s) = 1$$

8.46. Given that

$$G_c(s) = K \qquad G_p(s) = \frac{4s + 1}{s(s + 1)(s^2 + 2)} \qquad H(s) = 1$$

8.47. Given that

$$G_c(s) = K \qquad G_p(s) = \frac{4}{s(s + 1)} \qquad H(s) = 1$$

8.48. Given that

$$G_c(s) = K \qquad G_p(s) = \frac{5(s + 2)}{s^2 + 2s + 1} \qquad H(s) = 1$$

8.49. Given that

$$G_c(s) = K \qquad G_p(s) = \frac{10}{(s + 1)(s + 2)(s - 3)} \qquad H(s) = 1$$

8.50. Given that

$$G_c(s) = K \qquad G_p(s) = \frac{s^2 + 2s + 5}{s^4 + 3s^3 + 2s^2 + 4s + 2} \qquad H(s) = 1$$

8.51. Evaluate the system described in Exercise 8.41 for the case of positive feedback. Sketch the root-locus diagram.

8.52. Evaluate the system described in Exercise 8.42 for the case of positive feedback. Sketch the root-locus diagram.

8.53. Evaluate the system described in Exercise 8.43 for the case of positive feedback. Sketch the root-locus diagram.

8.54. Evaluate the system described in Exercise 8.44 for the case of positive feedback. Sketch the root-locus diagram.

8.55. Evaluate the system described in Exercise 8.45 for the case of positive feedback. Sketch the root-locus diagram.

8.56. Evaluate the system described in Exercise 8.46 for the case of positive feedback. Sketch the root-locus diagram.

8.57. Evaluate the system described in Exercise 8.47 for the case of positive feedback. Sketch the root-locus diagram.

8.58. Evaluate the system described in Exercise 8.48 for the case of positive feedback. Sketch the root-locus diagram.

8.59. Evaluate the system described in Exercise 8.49 for the case of positive feedback. Sketch the root-locus diagram.

8.60. Evaluate the system described in Exercise 8.50 for the case of positive feedback. Sketch the root-locus diagram.

8.61. Sketch the root loci for values of the controller gain K_c for the speed control system shown in Exercise 6.42, given that

$$K_e = 200, \quad K_t = 1, \quad \tau_e = 30 \text{ s}, \quad \tau_c = 3 \text{ s}$$

Also determine (1) The minimum settling time and (2) The minimum steady-state error that can be achieved with stable system behavior.

8.62. Consider the following closed-loop block diagram description of an armature-controlled dc motor (adapted from Dorf, 1974):

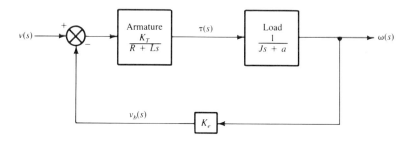

where $v(s)$, $\tau(s)$, and $\omega(s)$ represent the Laplace transforms of the armature voltage $v(t)$, the torque $\tau(t)$, and the angular velocity $\omega(s)$, respectively. In addition, the system parameters include the armature resistance R, the inductance L, the moment of inertia J of the load, the viscous damping coefficient a, the back emf $v_b(t)$, the torque constant K_T, and the voltage constant K_e. Assume zero initial conditions and

$$K_e = 0.2 \text{ V/rad/s}, \quad R = 0.5 \text{ } \Omega, \quad a = 0.1 \text{ in.-oz/rad/s}, \quad J = 0.2 \text{ oz-in.-s}^2,$$

$$L = 0.002 \text{ H}$$

(a) Develop and sketch the root-locus diagram for this system in terms of K_T.

(b) Identify the value of K_T that corresponds to critical damping.

9

ALTERNATIVE MODELING AND ANALYSIS SCHEMES

And here—ah, now, this really is something a little recherché.

Sherlock Holmes, The Adventure of The Musgrave Ritual
by Sir Arthur Conan Doyle

9.1 OBJECTIVES

Upon completion of this chapter, the reader should be able to:

- Identify the major characteristics of the alternative modeling schemes presented in this chapter.
- Define the frequency response of a system.
- Obtain the frequency response of a system, given the transfer function of the system and the applied sinusoidal forcing function.
- Obtain the z-transform of a function $c(t)$.
- Explain the application of z-transforms in the analysis of discrete or sampled-data systems.
- Identify some of the advantages and limitations of modern control system modeling.
- Obtain the state variables for simple systems via an appropriate change of variables.
- Explain the relationship between a point in state space and the state of a system.
- Recognize that linear, lumped-parameter models are limited but useful approximations of real nonlinear, continuous systems.
- Describe system compensation and relate it to the conflicting goals of stability and accuracy.

9.2 FREQUENCY RESPONSE

We have focused on the modeling and analysis of control systems in the Laplace or s domain; an alternative analytical method uses the *frequency response* of a system as its basis. The frequency response of a system is defined as its steady-state sinusoidal response to an applied sinusoidal input.

To develop the frequency response for a system, we begin with an applied forcing function of the form

$$r(t) = A \sin \omega t \tag{9.1}$$

with a corresponding Laplace transform given by

$$r(s) = \frac{A\omega}{s^2 + \omega^2} \tag{9.2}$$

Assume that the transfer function for the system is represented by $T(s)$ with the system output denoted by $c(s)$ in the Laplace domain. The output is then related to the input according to the expression

$$
\begin{aligned}
c(s) &= T(s)r(s) \\
&= \frac{T(s)A\omega}{s^2 + \omega^2} \\
&= \frac{T(s)A\omega}{(s + j\omega)(s - j\omega)}
\end{aligned} \tag{9.3}
$$

If we represent the roots of the characteristic equation for this system by $(r_1, r_2, \ldots, r_n)$, this response function can be expanded in terms of partial fractions:

$$c(s) = \frac{C_1}{s - j\omega} + \frac{C_2}{s + j\omega} + \frac{C_3}{s + r_1} + \frac{C_4}{s + r_2} + \cdots + \frac{C_{n+2}}{s + r_n} \tag{9.4}$$

The frequency response—defined as the steady-state system response—exists only if the system is stable. For a stable system, an inverse Laplace transformation produces

$$c(t) = C_1 e^{j\omega t} + C_2 e^{-j\omega t} + C_3 e^{r_1 t} + C_4 e^{r_2 t} + \cdots + C_{n+2} e^{r_n t} \tag{9.5}$$

which approaches

$$c(t) = C_1 e^{j\omega t} + C_2 e^{-j\omega t} \tag{9.6}$$

as $t \to \infty$. [If the system is stable, all of the roots $r_1, r_2, \ldots, r_n$ lie on the negative real half of the complex s-plane; all of the terms in equation (9.5) involving these roots then vanish as time t approaches infinity.] The coefficients C_1 and C_2 can be evaluated from equations (9.3) and (9.4); equating terms yields

$$\frac{T(s)A\omega}{(s+j\omega)(s-j\omega)} = \frac{C_1}{s-j\omega} + \frac{C_2}{s+j\omega} + \frac{C_3}{s+r_1} + \cdots \qquad (9.7)$$

We then multiply this relationship by $(s-j\omega)$ and set s equal to $j\omega$ in order to obtain an expression for C_1:

$$C_1 = \frac{T(j\omega)A}{2j} \qquad (9.8)$$

Similarly, multiplication of equation (9.7) by $(s+j\omega)$—followed by setting s equal to $-j\omega$—results in an expression for C_2:

$$C_2 = -\frac{T(-j\omega)A}{2j} \qquad (9.9)$$

$T(j\omega)$ is the transfer function for the system with the variable s replaced by the quantity $(j\omega)$; $T(s)$ is a complex number. Any complex number can be expressed in terms of its absolute value or *magnitude*, together with its *argument* (a polar angle; see Appendix A). For example, $T(j\omega)$ can be expressed as

$$T(j\omega) = |T(j\omega)|\, e^{j\phi} \qquad (9.10)$$

where $|T(j\omega)|$ represents the magnitude of $T(j\omega)$ and where ϕ is its argument, Similarly, we may write

$$T(-j\omega) = |T(j\omega)|\, e^{-j\phi} \qquad (9.11)$$

Equations (9.6) through (9.11) then give

$$c(t) = A|T(j\omega)|\frac{e^{j(\omega t+\phi)} - e^{-j(\omega t+\phi)}}{2j} \qquad (9.12)$$

However,

$$e^{j(\omega t+\phi)} = \cos(\omega t + \phi) + j\sin(\omega t + \phi) \qquad (9.13)$$

from which one can obtain the expression

$$\sin(\omega t + \phi) = \frac{e^{j(\omega t+\phi)} - e^{-j(\omega t+\phi)}}{2j} \qquad (9.14)$$

Therefore,

$$c(t) = A|T(j\omega)|\sin(\omega t + \phi) \qquad (9.15)$$

This is a significant result! The output $c(t)$ is the frequency response for the system. Equation (9.15) states that this response is sinusoidal in behavior, with a frequency identical to that of the input $r(t)$. Furthermore, the magnitude of the response is equal to the magnitude of the input multiplied by the absolute value of the modified transfer function $T(j\omega)$. Finally, there is a phase shift between the input and response signals equal to the argument ϕ of $T(j\omega)$.

In other words, if one simply replaces the variable s in the system's transfer function $T(s)$ by the quantity $j\omega$, the frequency response $c(t)$ of the system to a sinusoidal input can be easily obtained in terms of the magnitude and argument of the modified transfer function $T(j\omega)$, together with the applied input function $r(t)$.

In Section 6.3.3, we investigated the response of a first-order system to an applied sinusoidal function. The system's transfer function was given by

$$G(s) = \frac{1}{\tau s + 1}$$

whereas the applied forcing function was expressed in the form

$$f(t) = A \sin \omega t$$

Our results for the frequency response of the system [equation (9.15)] then predict that the output $x(t)$ for this system will be

$$x(t) = \frac{A}{(\tau^2 \omega^2 + 1)^{1/2}} \sin(\omega t + \phi) \tag{9.16}$$

(which is identical to our result in Section 6.3.3). To obtain equation (9.16), we simply replace the variable s in the expression for $G(s)$ by the quantity $j\omega$, followed by appropriate evaluation:

$$
\begin{aligned}
G(j\omega) &= \frac{1}{j\tau\omega + 1} \\[2mm]
&= \frac{1}{1 + j\tau\omega} \frac{1 - j\tau\omega}{1 - j\tau\omega} \\[2mm]
&= \frac{1 - j\tau\omega}{1 + \tau^2\omega^2} \\[2mm]
&= a + jb
\end{aligned}
\tag{9.17}
$$

where

$$a \equiv \frac{1}{1 + \tau^2\omega^2} \tag{9.18a}$$

$$b \equiv -\frac{\tau\omega}{1 + \tau^2\omega^2} \tag{9.18b}$$

As shown in Appendix A, the magnitude of a complex number $(a + jb)$ is given by $(a^2 + b^2)^{1/2}$. Furthermore, the argument of such a complex number is given by

$$
\begin{aligned}
\phi &= \tan^{-1}\frac{b}{a} \\[2mm]
&= \tan^{-1}\tau\omega
\end{aligned}
\tag{9.19}
$$

We then obtain

$$
\begin{aligned}
G(j\omega) &= |G(j\omega)|\, e^{j\phi} \\[1mm]
&= \left[\frac{(1)^2}{(1+\tau^2\omega^2)^2} + \frac{\tau^2\omega^2}{(1+\tau^2\omega^2)^2}\right]^{1/2} e^{j\phi} \\[2mm]
&= \frac{1}{(1+\tau^2\omega^2)^{1/2}}\, e^{j\phi}
\end{aligned}
\tag{9.20}
$$

Equation (9.16) is then directly obtained from expressions (9.15) and (9.20).

Expression (9.15) allows us to determine quickly the frequency response of any stable system for which the transfer function is known, as demonstrated above for the simple first-order system.

As one might expect for the analysis of systems to which are applied sinusoidal inputs, the frequency response method is very popular in electrical engineering applications. In addition, this method allows one to perform experiments on systems for which the transfer function is not known. The frequency response of a system to a controlled (well-defined) sinusoidal input can be investigated through such experimentation; given expressions for the known input and experimentally determined output, one can then formulate an approximate expression for the system's transfer function. Rapidly varying forcing functions (and disturbances to the system) can often be expressed as periodic functions, in which case the frequency response method is a very attractive approach in the analysis of a system (see Hale, 1973).

9.3 SAMPLED-DATA SYSTEMS AND THE z-TRANSFORM

Our presentation has dealt exclusively with *continuous* system inputs and outputs. With *discrete* or *sampled-data* systems, one deals with input and output information in the form of discrete pulses or sampling signals. One then seeks to use these sampled data to represent the corresponding continuous signal in the modeling, analysis, and control of the system.

If we define the quantities

$T \equiv$ time period between sampled units or pulses of data

$c_s(t) \equiv$ sampled signal

$c(t) \equiv$ continuous signal which is represented by the sampled signal $c_s(t)$

we then need to develop a mathematical relationship between the sampled signal $c_s(t)$ and the continuous signal $c(t)$. Such a relationship then allows us to analyze $c(t)$, given $c_s(t)$.

To develop the mathematical model of the sampled-data system, we first assume that $T \gg d$, where d is the duration time of the sampled pulse. As a result, the set of pulses can be represented by a corresponding set of impulses, where each impulse has an area equal to $c(nT)$ at any time $t = nT$ (see Hale, 1973; for a description of an impulse function, review Section 6.3.2 of this text.) Then

$$c_s(t) = c(0)\,\delta(t) + c(T)\,\delta(t - T) + \cdots + c(nT)\,\delta(t - nT) + \cdots \quad (9.21)$$

or

$$c_s(t) = \sum_{n=0}^{\infty} c(nT)\,\delta(t - nT) \quad (9.22)$$

In Section 3.6 we introduced the shift theorem, which stated that if the Laplace transform $f(s)$ of a function $F(t)$ is known, one can obtain the Laplace transform of the quantity $e^{at}F(t)$ in accordance with the relation

$$f(s - a) = \mathcal{L}[e^{at}F(t)] \quad (3.25)$$

This is also known as the *first shifting theorem* (Hale, 1973). A corresponding relationship between the function $e^{-as}f(s + a)$ in the Laplace domain and $F(t)$ can be expressed in the form

$$e^{-as}f(s + a) = \mathcal{L}[F(t)u(t - a)] \quad (9.23)$$

where $u(t)$ is the unit step function; equation (9.23) summarizes the *second shifting theorem.* Applying equation (9.23) to expression (9.22), we then obtain the Laplace transform of the sampled signal $c_s(t)$:

$$c_s(s) \equiv \mathcal{L}[c_s(t)]$$

$$= \sum_{0}^{\infty} c(nT)\,e^{-nTs} \quad (9.24)$$

If we then introduce a change of variable defined by

$$z \equiv e^{Ts} \quad (9.25)$$

we may write equation (9.24) in the form

$$c_s(s) = \sum_{0}^{\infty} c(nT)z^{-n} \quad (9.26)$$

Equation (9.26) is both the Laplace transform of $c_s(t)$ and the *z-transform* of $c(t)$, that is,

$$c(z) = \sum_{0}^{\infty} c(nT)z^{-n} \quad (9.27)$$

We may now perform a direct transformation of the continuous signal to the "z-domain" without constructing an intermediate Laplace transform of this signal. *z*-Transforms allow one to investigate the effect of sampling on a system. Furthermore, one can develop z-transfer functions for system components and perform an analysis of the stability of a sampled-data system in the z-plane. The z-transform allows us to convert transcendental [i.e., nonalgebraic; see Thomas (1960)] functions in s into algebraic functions in z, which can then be manipulated more easily than the original functions in s. An inverse z-transformation then provides information about the sampled signal behavior in the time domain.

Details about the use of z-transforms and sampled-data systems in control applications can be found in many excellent references [e.g., Kuo (1980), Kuo (1982), Cadzow (1973), Cadzow and Martens (1970), Hale (1973), Leigh (1985), Phillips and Nagle (1984)].

9.4 SYSTEM COMPENSATION

Recall that for a given forcing function, steady-state error e_{ss} may be reduced within a system by increasing the number n of open-loop poles located at the origin of the s-plane [i.e., by increasing the type number of the system (see Section 7.2)]. However, an increase in type number can lead to instability. As a result, the control engineer must seek to maintain stability while simultaneously increasing system accuracy—a difficult assignment!

System compensation allows the system designer to increase the number of poles and/or zeros within a system's transfer function through the introduction of appropriate components at strategic locations. *Series compensation* corresponds to the addition of poles/zeros in the feedforward portion of the system, whereas *parallel compensation* denotes the addition of poles/zeros in the feedback portion of a closed-loop path. *Series-parallel compensation* results from the addition of poles/zeros to both the feedforward and feedback portions of a system loop.

As one develops a suitable compensation scheme for a system, he or she often seeks to identify the dominant roots of the system's characteristic equation. One may then focus on these roots (nearest to the imaginary axis of the complex s-plane and located on the negative real half of this plane if the system is to remain stable) during the design of an appropriately compensated system, thereby decreasing the difficulties associated with such an effort.

Compensators are used to adjust the effect of controllers to achieve an overall desired system behavior. They form an integral part of the design effort for control systems [for additional information, see Harrison and Bollinger (1963), Hale (1973), Schwarzenbach and Gill (1978), Kuo (1982), and Palm (1983)].

9.5 NONLINEARITIES

Throughout our treatment, we have used linear, lumped-parameter models to represent real systems. Linear systems obey the principle of superposition and the principle of homogeneity (see Section 2.2). *Lumped-parameter* models represent all contributions to a particular system characteristic by a single ("lumped") value; for example, the resistance in an electrical circuit (which may be due to several sources) is represented by a single resistor.

Actual systems are nonlinear in behavior; we may adequately represent their behavior by linear models within *limited ranges* of operation (e.g., a spring may be represented by a linear approximation if it undergoes minimal extension or compression). If system components cannot be adequately described by linear approximations, special techniques must be used to analyze nonlinear behavior. These techniques are beyond the scope of our presentation; see, for example, Kreysig (1972), Hale (1973), Lefschetz (1965), Palm (1983), Vidyasagar (1978), and Minorsky (1969).

9.6 MODERN CONTROL THEORY AND STATE-SPACE MODELING

Modern control theory uses matrix vector equations to represent systems in an elegant and powerful manner. Hale (1973) identifies the myriad advantages of modern control system modeling, including:

- The ability to analyze nonlinear controllers which may be included within the system
- Replacing a large number of transfer functions used in the classical model of a complex linear system by a single matrix vector equation
- Automatic inclusion of initial conditions

As Hale also notes, there are practical limitations in the use of modern control system modeling; for example:

- Solutions may not always exist for all models.
- Analysis usually requires the use of a digital computer to generate numerical solutions.
- One needs to specify numerical values for system parameters in an analysis; as a result, numerous analyses by computer may be necessary to determine the effect of parameter modifications.

The professional control engineer should be familiar with both classical and modern control modeling techniques. We have focused on the classical approach in this text because it provides an introduction to control theory which emphasizes both clarity in analysis and insight into the behavior of physical systems.

Modern control theory is based on state-space modeling. *State space* is defined as an n-dimensional space in which n state variables act as coordinates; any point in state space—defined by a set of n simultaneous values for the n state variables—corresponds to a particular state of the system under consideration. The number n of state variables is the minimum number of linearly independent coordinates in state space which must be known in order to specify the state of the system. If the values of these n state variables are known at any given time t_0, the behavior of the system can be predicted for any time $t > t_0$.

The n state variables $(x_1, x_2, \ldots, x_n)$ form the coordinates of a *state vector* **x** in state space. We may similarly define a k-dimensional *control vector* vector **u** which represents the system inputs. Finally, a *function vector* **f** can then be formulated to represent the mathematical model of the system. This function vector is $(n + k + 1)$-dimensional, reflecting its dependence on the n-dimensional state vector **x**, the k-dimensional control vector **u**, and the time t. The function vector can be expressed as

$$\dot{\mathbf{x}}(t) = \mathbf{f}(x, u, t) = \frac{d\mathbf{x}}{dt} \tag{9.28}$$

or, equivalently, in the form of n first-order equations:

$$\begin{aligned}
\dot{x}_1 &= f_1(x_1, x_2, \ldots, x_n \,; u_1, \ldots, u_k \,; t) \\
\dot{x}_2 &= f_2(x_1, x_2, \ldots, x_n \,; u_1, \ldots, u_k \,; t) \\
&\ \ \vdots \qquad \vdots \\
\dot{x}_n &= f_n(x_1, x_2, \ldots, x_n \,; u_1, \ldots, u_k \,; t)
\end{aligned} \tag{9.29}$$

(Hale, 1973).

One may obtain these n first-order equations from a single nth-order differential equation through a simple change of variables; for example, the single nth-order differential equation

$$a_n \frac{d^n x}{dt^n} + a_{n-1} \frac{d^{n-1} x}{dt^{n-1}} + \cdots + a_1 \frac{dx}{dt} + a_0 x = u(t) \tag{9.30}$$

can be expressed as a set of n first-order equations if one introduces the following change of variables:

$$\begin{aligned}
x_1 &\equiv x \\
x_2 &\equiv \dot{x} = \frac{dx}{dt} \\
&\ \ \vdots \qquad \vdots \\
x_n &\equiv \frac{d^n x}{dt^n}
\end{aligned} \tag{9.31}$$

One such first-order equation would then be

$$\dot{x}_n = \frac{u(t)}{a_n} - \frac{a_{n-1}}{a_n} x_{n-1} - \frac{a_{n-2}}{a_n} x_{n-2} - \cdots - \frac{a_0}{a_n} x_1 \qquad (9.32)$$

The other $(n - 1)$ first-order equations can be directly written in a similar manner. The full set of n first-order equations then provide the n coordinates of the function vector, in accordance with equations (9.28) and (9.29).

A set of state variables that forms the mathematical model for a system is *not unique*; other sets may be formulated which also specify the state behavior of the system. The analyst seeks that set of state variables which simplifies the system analysis. Any feasible set of state variables $x_1, x_2, \ldots, x_n$ must contain only *linearly independent* x_i, that is, the relationship

$$\alpha_1 x_1 + \alpha_2 x_2 + \cdots + \alpha_n x_n = 0 \qquad (9.33)$$

must not be satisfied by any set of constants α_i other than zero (Hale, 1973).

Optimal control theory, in which one uses modern control analysis and state-space modeling, allows us to determine the system configuration or design that will optimize system performance with respect to selected criteria [e.g., minimum energy, minimum time; see Hale (1973)]. The reader interested in modern control theory and state-space modeling is encouraged to review the many excellent texts which focus on such analysis, for example, Ogata (1967), Hale (1973), Schwarzenbach and Gill (1978), Kuo (1982), Kailath (1980), Palm (1983), and Brogan (1985).

9.7 REVIEW

In summary, we have reviewed the following topics, facts, relationships, or concepts in this chapter.

- The *frequency response* of a system is defined as its steady-state sinusoidal response to an applied sinusoidal input. If the transfer function for a system is represented by $T(s)$, the frequency response $c(t)$ of the system can be expressed as

$$c(t) = A|T(j\omega)| \sin(\omega t + \phi) \qquad (9.15)$$

  where the input to the system is equal to $A \sin \omega t$. $T(j\omega)$ is obtained by replacing the variable s in the transfer function $T(s)$ by the quantity $j\omega$. The absolute value $|T(j\omega)|$ is equal to the ratio of the amplitudes of the system response $c(t)$ and the applied sinusoidal forcing function $r(t)$. The argument ϕ of $T(j\omega)$ is equal to the phase shift between the input and output signals.

- With *discrete* or *sampled-data* systems, one deals with input and output information in the form of discrete pulses or sampling signals. With the aid of the second shifting theorem and a change of variable given by the expression

$$z = e^{Ts} \qquad (9.25)$$

(where T represents the time period between pulses of data), we obtained the *z-transform* of $c(t)$:

$$c(z) = \sum_{0}^{\infty} c(nT)z^{-n} \qquad (9.27)$$

with which sampled-data systems may be analyzed.

- *Linear, lumped-parameter* models are limited but useful approximations of real systems which are continuous and *nonlinear* in behavior.

- *System compensation* allows the system designer to increase the number of poles and zeros within a system's transfer function through the introduction of appropriate components ("compensators") at strategic locations within the system. Such compensation allows one to increase system accuracy while maintaining stability.

- *Modern control theory* is based on state-space modeling. State space is defined as an n-dimensional space in which n *state variables* act as coordinates. Any point in state space corresponds to a particular state of the system under consideration. The n state variables form the coordinates of a state vector **x** in state space. A function vector **f**, equal to $\dot{x}$, can be developed through a change of variables in which a single nth-order differential equation (for example) is replaced by n first-order equations. The full set of n first-order equations then provides the n coordinates of the function vector, in accordance with equations (9.28) and (9.29).

- A set of state variables that forms the mathematical model for a system is not unique; other sets may be formulated which also specify the state behavior of the system. Furthermore, all the state variables must be linearly independent.

EXERCISES

9.1. Determine the frequency response of the system shown in Exercise 6.1 to an applied forcing function given by

$$f(t) = 2 \sin 3t$$

9.2. Determine the frequency response of the system shown in Exercise 6.39 to an applied forcing function given by

$$f(t) = 5 \sin 2t$$

9.3. Determine the frequency response of the system shown in Exercise 6.40 to an applied forcing function given by

$$f(t) = 3 \sin 4t$$

9.4. Determine the frequency response of the system shown in Exercise 6.41 to an applied forcing function given by

$$f(t) = 2 \sin 3t$$

9.5. Determine the frequency response of the system shown in Exercise 7.3 to an applied forcing function given by

$$f(t) = 4 \sin 2t$$

9.6. Determine the frequency response of the system shown in Exercise 7.33 to an applied forcing function given by

$$f(t) = 2 \sin 3t$$

9.7. Determine the frequency response of the system shown in Exercise 8.1 to an applied forcing function given by

$$f(t) = 3 \sin 6t$$

9.8. Determine the frequency response of the system shown in Exercise 8.41 to an applied forcing function given by

$$f(t) = 3 \sin 7t$$

9.9. Determine the z-transform of the unit step function $f(t)$.

9.10. Identify five different systems which can be classified as discrete or sampled-data systems.

9.11. Determine the state variables for the system shown in Exercise 6.1. (Recall that a set of state variables is not unique.) Demonstrate that the set of state variables which has been chosen by you obeys the requirement of linear independence.

9.12. Determine a set of state variables for the system shown in Exercise 6.5. Show that these state variables are linearly independent.

APPENDICES

*I shall take nothing for granted until I have
the opportunity of looking personally into it.*

Sherlock Holmes, The Boscombe Valley Mystery
by Sir Arthur Conan Doyle

A: COMPLEX VARIABLES

A complex variable z can be expressed as the sum of a purely real quantity
(x) and an imaginary quantity (jy), where j represents the square root of -1:

$$z = x + jy \tag{A.1}$$

An equivalent expression for z is

$$z = r \cos \theta + jr \sin \theta \tag{A.2}$$

where r is the *absolute value* of the complex number z, given by

$$r = (x^2 + y^2)^{1/2} \tag{A.3}$$

and where the polar angle θ is the *argument* of z; the *Argand diagram* shown
in Figure A.1 presents the graphical interpretation of these quantities in the
complex plane formed by the real and imaginary axes.

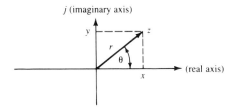

Figure A.1 Argand diagram.

The complex conjugate (z^*) of z is expressed in the form

$$z^* = x - jy \qquad (A.4)$$

The complex conjugate z^* allows us to express r as

$$r = (z \cdot z^*)^{1/2}$$
$$= [(x + jy)(x - jy)]^{1/2} \qquad (A.5)$$
$$= (x^2 + y^2)^{1/2}$$

Furthermore, the ratio of two complex numbers can be simplified through the use of complex conjugates:

$$\frac{a + jb}{c + jd} = \frac{(a + jb)(c - jd)}{(c + jd)(c - jd)}$$
$$= \frac{ac + j(bc) - j(ad) - j^2(ad)}{c^2 + d^2} \qquad (A.6)$$
$$= \frac{ac + bd}{c^2 + d^2} + j\frac{bc - ad}{c^2 + d^2}$$

that is, we obtain two distinct terms—one that is purely real and one that is the imaginary portion of the result.

Finally (with r set equal to 1) *de Moivre's theorem* can be very helpful when one is working in the complex plane (see Thomas, 1960):

$$(\cos\theta + j\sin\theta)^n = \cos n\theta + j\sin n\theta \qquad (A.7)$$

B: INTEGRATION BY PARTS

A general method of integration, known as integration by parts, is based on the formula for the differential of a product of two functions u and v (Thomas, 1960):

$$d(uv) = u\,dv + v\,du \qquad (B.1)$$

Integration and rearrangement of terms then produces the expression

$$\int u\,dv = uv - \int v\,du + C \qquad (3.12)$$

where C is a constant of integration.

As an example, consider the integral

$$\int x^2\,e^{4x}\,dx = ? \qquad (B.2)$$

If we identify terms as follows:

$$u \equiv x^2 \qquad \rightarrow \qquad du = 2x \, dx$$

$$dv \equiv e^{4x} \, dx \qquad \rightarrow \qquad v = \tfrac{1}{4} e^{4x}$$

we obtain with equation (3.12),

$$\int x^2 \, e^{4x} \, dx = \tfrac{1}{4} x^2 \, e^{4x} - \tfrac{1}{2} \int e^{4x} x \, dx$$

Use of the expression (3.12) once again will complete our analysis; we define

$$u_2 \equiv x \qquad \rightarrow \qquad du_2 = dx$$

$$dv_2 \equiv e^{4x} \, dx \qquad \rightarrow \qquad v_2 = \tfrac{1}{4} e^{4x}$$

Therefore,

$$\int e^{4x} x \, dx = \tfrac{1}{4} x e^{4x} - \tfrac{1}{16} e^{4x}$$

Finally, we obtain

$$\int x^2 e^{4x} \, dx = e^{4x} (\tfrac{1}{4} x^2 - \tfrac{1}{8} x + \tfrac{1}{32}) + C$$

As another example, consider the integral

$$\int x \sin 5x \, dx = ? \qquad \qquad (B.3)$$

We define the quantities

$$u \equiv x \qquad \rightarrow \qquad du = dx$$

$$dv \equiv \sin 5x \, dx \qquad \rightarrow \qquad v = -\tfrac{1}{5} \cos 5x$$

so that we obtain

$$\int x \sin 5x \, dx = -\frac{x}{5} \cos 5x + \tfrac{1}{25} \sin 5x + C$$

C: METHOD OF PARTIAL FRACTIONS

The method of partial fractions allows us to simplify the process of performing inverse Laplace transformations by rewriting a complex fraction as an equivalent sum of fractions which have simpler denominators.

Consider the following fraction and an equivalent sum of partial fractions:

$$\frac{K}{(x - r_1)(x - r_2)(x - r_3)} = \frac{A}{x - r_1} + \frac{B}{x - r_2} + \frac{C}{x - r_3} \qquad (C.1)$$

where r_1, r_2, and r_3 represent roots of the third-order denominator of the fraction given on the left side of the equality. A, B, and C are undetermined coefficients which must now be found. To determine the value of A, one simply multiplies both sides of the equality by the denominator $(x - r_1)$ associated with the coefficient A, after which x is set equal to r_1. One then obtains

$$\frac{K(x - r_1)}{(x - r_1)(x - r_2)(x - r_3)} = A + \frac{B(x - r_1)}{x - r_2} + \frac{C(x - r_1)}{x - r_3}$$

which becomes, upon the substitution $x = r_1$,

$$\frac{K}{(r_1 - r_2)(r_1 - r_3)} = A + 0 + 0$$

Similarly, one may calculate the values for B and C.

As an example, consider the fraction and partial fractions given below:

$$\frac{5}{(x - 2)(x - 3)(x - 4)} = \frac{A}{x - 2} + \frac{B}{x - 3} + \frac{C}{x - 4} \qquad (C.2)$$

To determine the value of A, we multiply both sides of this equality by $(x - 2)$ and then set x equal to 2:

$$\frac{5}{(x - 3)(x - 4)} = \frac{5}{(2 - 3)(2 - 4)} = \frac{5}{2} = A$$

For B, we multiply equation (C.2) by the term $(x - 3)$ and then set $x = 3$:

$$\frac{5}{(x - 2)(x - 4)} = \frac{5}{(3 - 2)(3 - 4)} = -5 = B$$

Finally, C can be obtained by multiplying equation (C.2) by $(x - 4)$ and then setting $x = 4$:

$$\frac{5}{(x - 2)(x - 3)} = \frac{5}{(4 - 2)(4 - 3)} = \frac{5}{2} = C$$

If the original denominator has a factor $(x - r_1)^m$—that is, there are m roots of the original denominator equal to r_1—one may multiply the equation (C.2) by this factor, collect terms, and identify equivalent coefficients on each side of the equality. Furthermore, if the original fraction's numerator is a function $f(x)$ (but not a constant), where the degree of $f(x)$ is less than that of the fraction's denominator, one may express the original fraction as a sum of partial fractions in which quadratic factors $(x^2 + px + q)$ of the original denominator compose the denominators of the partial fractions. For example, if the original fraction is the ratio of $f(x)$ to $g(x)$, and if $g(x)$ has a quadratic factor $(x^2 + px + q)$ which divides $g(x)$ n times, a contribution to the total sum of partial fractions due to these quadratic factors is given by

$$\frac{B_1 x + C_1}{x^2 + px + q} + \frac{B_2 x + C_2}{(x^2 + px + q)^2} + \cdots + \frac{B_n x + C_n}{(x^2 + px + q)^n}$$

Each set of quadratic factors will produce a similar set of terms in the expansion of partial fractions.

As an example, consider the fraction and equivalent sum of partial fractions given below:

$$\frac{25x + 4}{(x^2 + 4)(x - 2)} = \frac{Bx + C}{x^2 + 4} + \frac{A}{x - 2}$$

Multiplying this equality by the factors $(x^2 + 4)$ and $(x - 2)$, we have

$$25x + 4 = (Bx + C)(x - 2) + A(x^2 + 4)$$

$$= (B + A)x^2 + (C - 2B)x + (4A - 2C)$$

Equating coefficients of similar terms in x on both sides of this equality then produces the following relations:

$$0 = B + A$$

$$25 = C - 2B$$

$$4 = 4A - 2C$$

We may then solve these expressions for A, B, and C:

$$A = \tfrac{54}{8}$$

$$B = -\tfrac{54}{8}$$

$$C = \tfrac{92}{8}$$

(Further details can be found in Thomas, 1960.)

D: BOND GRAPHING AS A MODE OF TECHNICAL COMMUNICATION*

Abstract—Bond graphs offer a simple, efficient method for developing models of physical (e.g., mechanical, electrical, thermal, fluid), chemical, economic, and other types of systems. They

* "Bond Graphing as a Mode of Technical Communication," by Gerard Voland, from *IEEE Transactions on Professional Communication*, Volume PC-25, Number 1, March 1982, pp. 35-37, copyright © 1982 by The Institute of Electrical and Electronics Engineers, Inc., known as the IEEE.

are a communication device that can be easily interpreted by workers with differing technical and mathematical backgrounds, thereby facilitating interdisciplinary discussion and exchange of information. This paper provides a quick introduction to bond graphs with examples and references.

Schematics, block diagrams, flowcharts, tables, etc., allow engineers to represent data, relationships between system components and system variables, and a variety of system models in an informative yet simple manner. Another graphic form of physical modeling has been undergoing development since its creation in 1959 [1]: *bond graphing*. This is a graphic representation in which the number of defined variables is small and which can be easily interpreted by workers with differing technical and mathematical backgrounds, thereby increasing communication between specialists in various fields. Although bond-graph modeling has been used to describe physical, chemical, economic, and other types of systems during the past two decades (as evidenced by the interdisciplinary bibliography of Gebben [2]), it remains an unfamiliar method of communication to most technical workers.

Bond graphs provide a simple, shorthand notation for differential equations that relate system variables. In addition, the organizational structure of the system is contained within the graph and a single graph can be used to represent all of the energy domains (e.g., thermal, fluid, mechanical, electrical) of a system. This modeling technique can thus be used by specialists in a broad range of disciplines to identify and represent the significant similarities of the systems with which they work.

Bond Graph Modeling

With bond graphs, a system is described by a set of basic elements that either store or dissipate energy. Some of these elements are shown in Table D.1 with examples of physical components that can be represented. Bonds between the elements allow power transmission to occur in a variety of forms. This power flow is described by a set of variables,

$e \equiv$ effort variable (e.g., force, torque, pressure, voltage)

$f \equiv$ flow variable (e.g., velocity, angular velocity, volume flow rate, current)

such that

$$P = e \cdot f \equiv \text{instantaneous power}$$

The direction of power flow is indicated by a half-arrow representation of the bond, i.e., $\rightharpoonup$. Finally, a "causal stroke" | is used to indicate the direction of the effort-variable signal, i.e., if two elements A and B are bonded according to

$$A \xrightarrow[f]{e} | B$$

TABLE D.1 Basic Bond Graph Modeling Elements

Bond graph element	Symbol	Physical element		
Resistor, R	$\dfrac{e}{f}\,R$	$\dfrac{V}{i}\,R$ Electrical resistor	$\dfrac{F}{\dot{x}}\,R$ Mechanical dashpot	$P = P_1 - P_2$ $\dfrac{P}{v}\,R$ Porous (fluid) plug
Capacitor, C	$\dfrac{e}{f}\,C$	$\dfrac{V}{i}\,C$ Capacitor	$\dfrac{F}{x}\,C$ Spring	$\dfrac{P}{v}\,C$ Gravity tank
Inertia, I	$\dfrac{e = p}{f}\,I$	$\dfrac{V}{i}\,I$ Inductance	$\dfrac{F}{\dot{x}}\,I$	$(P = P_1 - P_2)$ $\dfrac{P}{v}\,I$

Mechanical and fluid inertia effects

then the effort variable e is an input signal to element B, whereas the flow variable f is an input signal to A (and an output signal from B). In block-diagram notation the equivalent is

$$A \rightleftharpoons B$$

Existing computer software packages, known collectively as ENPORT [3], allow one to determine the equations, system eigenvalues, integration step sizes, and transient response of all state variables for linear systems modeled with bond graphs.

Examples

Figure D.1(a) shows a mass-spring system; the corresponding bond graph model is developed in Figure D.1(b). The relationships between the system variables are all contained within the bond graph structure; they are given mathematically in Table D.2 for comparison.

An example of fluid-flow modeling in Figure D.2(a) shows blood flow through perfused tissue [4, 5]. Fluid resistance and compliance are included in the corresponding bond graph model shown in Figure D.2(b). (Additional characteristics of this physical system need to be included in the graphic model for it to be an accurate representation; see Voland [6] for a more sophisticated bond graph model of a circulation process.)

Karnopp and Rosenberg [7] provide a much more detailed introduction to bond graph modeling techniques.

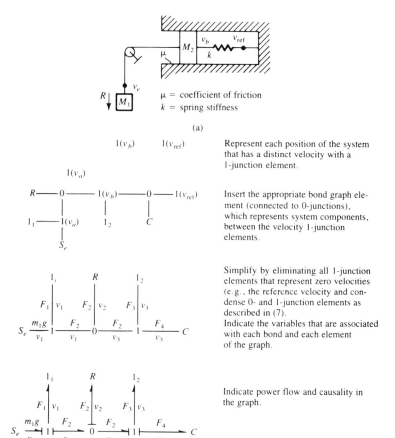

Figure D.1 (a) Mechanical mass-spring system. (b) Development of a bond graph for the mass-spring system. F_i and $m_1 g$ are forces, M_l are masses, v_i are velocities, and S_e is the effort source.

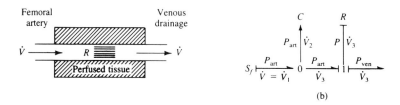

Figure D.2. (a) Blood flow through tissue [4, 5]. (b) Bond graph for the blood-flow system. Net pressure $P = P_{art} - P_{ven}$. V_i are volume flow rates, and S_f is the flow source.

**TABLE D.2   Mathematical Relationships
between Variables of the
Mass-Spring System**

1-junction	$F_1 = m_1 g - F_2$
I_1	$v_1 = \dfrac{1}{M_1} \displaystyle\int_0^t F_1 \, dt$
0-junction	$v_2 = v_1 - v_3$
R	$F_2 = \mu v_2$
1-junction	$F_3 = F_2 - F_4$
I_2	$v_3 = \dfrac{1}{M_2} \displaystyle\int_0^t F_3 \, dt$
C	$F_4 = k \displaystyle\int_0^t v_3 \, dt$

ACKNOWLEDGMENT

This work [Appendix D] was supported in part by grant RR07143 from the U.S. Department of Health and Human Services.

REFERENCES

1. H. M. Paynter, *Analysis and Design of Engineering Systems*, Cambridge, MA: M.I.T. Press, 1961.

2. V. D. Gebben, "Bond Graph Bibliography for 1961-1976," *Journal of Dynamic Systems, Measurement and Control, Trans. ASME, Series G*, vol. 99, pp. 143-145, 1977.

3. R. C. Rosenberg, *A User's Guide to ENPORT-4*. New York: John Wiley, 1974.

4. J. M. Ross, H. M. Fairchild, J. Weldy, and A. C. Guyton, "Autocoregulation of Blood Flow by Oxygen Lack," *American Journal of Physiology*, vol. 202, pp. 21-24, 1962.

5. D. S. Riggs, *Control Theory and Physiological Feedback Mechanisms*. Huntington, NY: R. E. Krieger, 1976.

6. G. Voland, "Bond Graph Modeling Applied to the Human Heart," accepted for publication in *Engineering Design Graphics Journal*, 1982.

7. D. C. Karnopp and R. C. Rosenberg, *System Dynamics: A Unified Approach*. New York: John Wiley, 1975.

REFERENCES

ARDEN, B. W., and ASTILL, K. N., *Numerical Algorithms: Origins and Applications*, Addison-Wesley Publishing Company, Inc., Reading, Mass. (1970).

BEER, F. P., and JOHNSTON, E. R., JR., *Mechanics for Engineers: Statics and Dynamics*, 3rd ed., McGraw-Hill Book Company, New York (1976).

BENDER, C. M., and ORSZAG, S. A., *Advanced Mathematical Methods for Scientists and Engineers*, McGraw-Hill Book Company, New York (1978).

BENNETT, S., *A History of Control Engineering 1800-1930*, Institution of Electrical Engineers, London (1979).

BROGAN, W, L., *Modern Control Theory*, 2nd ed., Prentice-Hall, Inc., Englewood Cliffs, N.J. (1985).

CADZOW, J. A., *Discrete Time Systems: An Introduction with Interdisciplinary Applications*, Prentice-Hall, Inc., Englewood Cliffs, N.J. (1973).

CADZOW, J. A., and MARTENS, H. R., *Discrete Time and Computer Control Systems*, Prentice-Hall, Inc., Englewood Cliffs, N.J. (1970).

CARNAHAN, B., LUTHER, H. A., and WILKES, J. O., *Applied Numerical Methods*, John Wiley & Sons, Inc., New York (1969).

CHIN, S. L., "A Mathematical Model of Human Productivity versus Encouragement," *IEEE Transactions on Systems, Man and Cybernetics*, Vol. SMC-14, No. 2, pp. 302-304, March-April (1984).

CHURCHILL, R. V., *Operational Mathematics*, McGraw-Hill Book Company, New York (1958).

COUGHANOWR, D. R., and KOPPEL, L. B., *Process Systems Analysis and Control*, McGraw-Hill Book Company, New York (1965).

DEO, N., *System Simulation with Digital Computer*, Prentice-Hall, Inc., Englewood Cliffs, N.J. (1983).

DOEBELIN, E. O., *Dynamic Analysis and Feedback Control*, McGraw-Hill Book Company, New York (1962).

DORF, R. C., *Modern Control Systems*, 2nd ed., Addison-Wesley Publishing Company, Inc., Reading, Mass. (1974).

EVANS, W. R., "Graphical Analysis of Control Systems," *Transactions of the AIEE*, Vol. 67, pp. 547–551 (1948).

EVANS, W. R., "Control System Synthesis by Root Locus Method," *Transactions of the AIEE*, Vol. 69, pp. 1–4 (1950).

EVANS, W. R., *Control System Dynamics*, McGraw-Hill Book Company, New York (1954).

FORWARD, R. L., "Electronic Damping of Orthogonal Bending Modes in a Cylindrical Mast—Experiment," *Journal of Spacecraft and Rockets*, Vol. 18, No. 1, pp. 11–17, January–February (1981).

GORDON, G., *System Simulation*, 2nd ed., Prentice-Hall, Inc., Englewood Cliffs, N.J. (1978).

HALE, F. J., *Introduction to Control System Analysis and Design*, Prentice-Hall, Inc., Englewood Cliffs, N.J. (1973).

HARRISON, H. L., and BOLLINGER, J. G., *Introduction to Automatic Controls*, International Textbook Company, Scranton, Pa. (1963).

HENRICI, P., *Discrete Variable Methods in Ordinary Differential Equations*, John Wiley & Sons, Inc., New York (1962).

HOLL, W. H., "Design Analysis of an Automotive Speed Control System," *Proceedings, JACC*, p. 45 (1963).

HORI, H., KITAYAMA, Y., KITANO, M., YABUZAKI, T., and OGAWA, T., "Frequency Stabilization of GaAlAs Laser Using a Doppler-Free Spectrum of the Cs-D_2 Line," *IEEE Journal of Quantum Electronics*, Vol. QE-19, No. 2, pp. 169–175, February (1983).

HUNTER, R. P., *Automated Process Control Systems: Concepts and Hardware*, Prentice-Hall, Inc., Englewood Cliffs, N.J. (1978).

JACOBY, S. L. S., and KOWALIK, J. S., *Mathematical Modeling with Computers*, Prentice-Hall, Inc., Englewood Cliffs, N.J. (1980).

KAILATH, T., *Linear Systems*, Prentice-Hall, Inc., Englewood Cliffs, N.J. (1980).

KAY, J. M., *An Introduction to Fluid Mechanics and Heat Transfer*, 2nd ed., Cambridge University Press, London (1963).

KREYSIG, N., *Advanced Engineering Mathematics*, 3rd ed., John Wiley & Sons, Inc., New York (1972).

KUO, B. C., *Automatic Control Systems*, 3rd ed. (1975) and 4th ed. (1982), Prentice-Hall, Inc., Englewood Cliffs, N.J.

KUO, B. C., *Digital Control Systems*, Holt, Rinehart and Winston, New York (1980).

LEFSCHETZ, S., *Stability of Nonlinear Control Systems*, Academic Press, Inc., New York (1965).

LEIGH, J. R., *Applied Digital Control: Theory, Design and Implementation*, Prentice-Hall, Inc., Englewood Cliffs, N.J. (1985).

LENK, J. D., *Handbook of Controls and Instrumentation*, Prentice-Hall, Inc., Englewood Cliffs, N.J. (1980).

LENK, J. D., *Handbook of Microcomputer-based Instrumentation and Controls*, Prentice-Hall, Inc., Englewood Cliffs, N.J. (1984).

LUH, J. Y. S., "Conventional Controller Design for Industrial Robots—A Tutorial," *IEEE Transactions on Systems, Man and Cybernetics*, Vol. SMC-13, No. 3, pp. 298–316, May–June (1983).

MARION, J. B., *Physics: The Foundations of Modern Science*, John Wiley & Sons, Inc., New York (1973).

MAYR, O., *The Origins of Feedback Control* (English translation), The MIT Press, Cambridge, Mass. (1970).

MINORSKY, N., *Theory of Nonlinear Control Systems*, McGraw-Hill Book Company, New York (1969).

OGATA, K., *State Space Analysis of Control Systems*, Prentice-Hall, Inc., Englewood Cliffs, N.J. (1967).

OGATA, K., *Modern Control Engineering*, Prentice-Hall, Inc., Englewood Cliffs, N.J. (1970).

OGATA, K., *System Dynamics*, Prentice-Hall, Inc., Englewood Cliffs, N.J. (1978).

O'MEARA, T. R., "The Multidither Principle in Adaptive Optics," *Journal of the Optical Society of America*, Vol. 67, No. 3, pp. 306–315, March (1977).

PALM, W. J., III, *Modeling, Analysis and Control of Dynamics Systems*, John Wiley & Sons, Inc., New York (1983).

PEPPER, D. M., "Nonlinear Optical Phase Conjugation," *Optical Engineering*, Vol. 21, No. 2, pp. 156–183, March–April (1982).

PHILLIPS, C. L., and NAGLE, H. T., *Digital Control System Analysis and Design*, Prentice-Hall, Inc., Englewood Cliffs, N.J. (1984).

PUGH, A., III, *DYNAMO II User's Manual*, The MIT Press, Cambridge, Mass. (1973).

RAVEN, F. H., *Automatic Control Engineering*, 3rd ed., McGraw-Hill Book Company, New York (1978).

RICHARDSON, L. F., *Arms and Insecurity*, Borwood, Pittsburgh, Pa. (1960).

ROSE, R. M., SHEPARD, L. A., and WULFF, J., *The Structure and Properties of Materials*, Vol. IV: *Electronic Properties*, John Wiley & Sons, Inc., New York (1966).

ROUTH, E. J., *Dynamics of a System of Rigid Bodies*, 3rd ed., The Macmillan Company, New York (1877).

SAATY, T. L., and ALEXANDER, J. M., *Thinking with Models*, Pergamon Press Ltd., Oxford (1981).

SCHWARZENBACH, J., and GILL, K. F., *System Modeling and Control*, John Wiley & Sons, Inc., New York (1978).

SPECKHART, F. H., and GREEN, W. L., *A Guide to Using CSMP*, Prentice-Hall, Inc., Englewood Cliffs, N.J. (1976).

SWIGERT, C. J., and FORWARD, R. L., "Electronic Damping of Orthogonal Bending Modes in a Cylindrical Mast—Theory," *Journal of Spacecraft and Rockets*, Vol. 18, No. 1, pp. 5–10, January–February (1981).

THOMAS, G. B., JR., *Calculus and Analytic Geometry*, 3rd ed., Addison-Wesley Publishing Company, Inc., Reading, Mass. (1960).

TOLLE, M., "Beitrage zur Beurteilung der Zentrifugalpendelregulatoren," *Zeitschrift des Vereines deutscher Ingenieure*, Vol. 39, pp. 735–741, 776–779, 1492–1498, 1543–1551 (1895); Vol. 40, pp. 1424–1428, 1451–1455 (1896).

TUTSIM, *TUTSIM User's Manual for IBM PC Computer*, Twente University of Technology, Simulation Software, The Netherlands (North American Distributor: APPLIED i, Palo Alto, Calif.) (1984).

VIDYASAGAR, M., *Nonlinear Systems Analysis*, Prentice-Hall, Inc., Englewood Cliffs, N.J. (1978).

VOLAND, G., "Bond Graphing as a Mode of Technical Communication," *IEEE Transactions on Professional Communication*, Vol. PC-25, No. 1, pp. 35–37, March (1982).

VOLAND, G., and VOLAND, M., "Decisive Factors in Systems Modeling and Computer Simulation," *Computers in Education Journal*, Vol. 3, No. 6, pp. 8–9, November–December (1983a).

VOLAND, G., and VOLAND, M., "Dynamics of Systems Modeling and Simulation," *Computers in Engineering 1983*, Vol. 1: *Computer-Aided Design, Manufacturing and Simulation*, American Society of Mechanical Engineers, New York (1983b).

WARK, K., *Thermodynamics*, 3rd ed., McGraw-Hill Book Company, New York (1977).

WIENER, N., *Cybernetics*, John Wiley & Sons, Inc., New York (1948).

WILCOX, R. B., "Dynamic Analysis and Simulation of Management Control Functions," *Proceedings, Second IFAC Congress*, Vol. 2, Butterworth & Company Ltd., London (1963).

INDEX

A

Absolute instability, 5, 19
Accuracy, 4, 6, 7, 9, 179–210, 235
Across variables, 117, 144, 168
Actuating signal, 108, 180, 188
Actuators, 189
Analogous systems, 12, 111, 117, 140–56, 168
Angle criterion, 223, 229
Argand diagram, 247
Armament competition, 44, 138–40
Automatic control system, 2, 8
Auxiliary equation, 216–18, 229–30

B

Block diagrams, 2, 9, 47, 106–16, 167
Bond graphing, 113, 167, 251–55
Boundary-value problems, 89
Branch point, 109
Breakaway point, 224, 227, 230

C

Capacitive elements, 118, 140–41, 143, 168

Cauchy, Augustin-Louis, 49
Chain rule, 91
Characteristic equation, 11, 15, 38
Closed-loop systems, 1, 2, 9
Common coefficients, method of, 11, 32, 40
Compensation, 235, 241, 245
Complementary function, 14
Complex variables, 16–17, 238, 247–48
Compliance, 141
Continuous systems, 89, 235
Control actions, 179–210
    applications of, 195–200, 203
    basic principle for, 179, 200–1
Controller, 2
Control systems, 1, 8
Coulomb friction, 143
Cramer's rule, 11, 34, 40, 138
Critical damping, 117, 131, 136, 168, 221
CSMP, 167
Cybernetics, 4

D

Damping ratio, 117, 130, 156, 167

DC motor, 152–56
Decay ratio, 117, 138, 168
de Moivre's theorem, 248
Derivative control, 192–94
Difference operator, 108, 113
Differential equation, 11–45
Differential gap, 189
Differential mean-value theorem, 78
Differential operator, 15
Discrete systems, 235, 239–41, 245
Discretization error, 90
Disturbance, 3
Drebbel, Cornelis, 3
Dry friction, 142
Dynamic error, 162
DYNAMO, 167

E

Energy domains, 146, 168
Environment, 2–3
Equivalent open-loop systems, 109,
   113
Error, 108
Error constants, 182, 202
Error detector, 108
Euler's method, 73, 88–91, 100
Evans, W. R., 220
Exponential order, 49, 52

F

False positions, method of, 73, 78–80,
   99
Feedback, 1–2, 9
Feedback control, history of, 3
Feedback signal, 109, 188
Feedforward, 109
Final-value theorem, 47, 67–70, 182,
   198–99, 202, 211–12
Finite differences, 76, 78
First-order systems, 117, 119–20,
   167
Float-valve regulators, 3, 190
Force-current analogy, 145
Force-voltage analogy, 145
Forcing functions, 12
Frequency response, 235–39, 244
Friction, 142–43

G

Gain, 122, 156, 190
General first-order systems, 117, 120–29,
   167
General input, 130
Generalized equations, 118, 168
General linear integral transformation,
   46, 48, 68
General output, 130
General second-order systems, 117,
   129–40, 167, 195
Graphical methods, 12

H

Homogeneity, principle of, 13, 47, 242
Huen's method, 92, 100
Hurwitz determinants, 212

I

Ideal behavior, 142
Ideal gas, equation of state for, 82
Image of function, 47
Improved Euler's method, 92, 100
Improved Polygon method, 93, 100
Inductive elements, 117, 141, 143, 168
Initial-value problems, 89
Initial-value theorem, 47, 67–69
Integral control, 179, 182, 187–88,
   190–92, 202
Integral gain, 190
Integral time, 191
Integration by parts, 51–52, 248–49
Inverse Laplace transformation, 46,
   55–57, 69

K

Kernel of transformation, 48
Kinetic friction, 143
Kirchhoff's voltage law, 149–50
Ktesibius, 3, 9

L

Laplace transformation, 12, 46–72
   of derivatives, 52–55

sufficient conditions for existence of, 46

Lerch theorem, 56

Linear approximation, 11

Linearity, 13, 24

Linear transformation, 47

Lin's method for quadratic factors, 73, 80–88, 99

Lumped-parameter system modeling, 37, 89, 235, 242, 245

M

Magnitude criterion, 223, 229

Management-production system, 156–66

Manipulated signals, 188–89

Mapping of a function, 46, 55, 69

Marginal stability, 211, 216, 218, 228–30

Mechanical inductance, 141

Method of common coefficients, 11, 32, 40

Modern control theory, 235, 242–45

Modified Euler's method, 93, 100

Moment of inertia, 141

Multiple system outputs, 31, 40

N

Neutral zone, 189

Newton-Raphson method, 73, 75–77, 98

Newton's second law of motion, 129

Nonlinearities, 242, 245

Nonlinear ordinary differential equations, 73, 88–98

Nonlinear systems, 235, 242, 245

Null function, 55

Numerical approximation methods, 12, 73–105

O

On-off control, 179, 189, 202

Open-loop systems, 1–2, 9

Operational mathematics, 47

Operator notation, 14

Optimal control theory, 244

Ordinary differential equations, 12

Overdamping, 117, 131, 136–37, 168, 221

Overshoot, 117, 137, 157, 168, 193, 198, 220

P

Partial differential equations, 12, 89

Partial fractions, 58–59, 61, 65, 196, 236, 249–51

Particular integral, 14, 20

Poles, 223–24, 226, 230

Predator-prey system, 219

Predictor-corrector method, 93, 100

Productivity, 123, 126, 156–66

Proportional control, 179, 189–90, 192–93, 202, 220

Proportional-plus-derivative control, 179, 192–94, 202

Proportional-plus-derivative-plus-integral control, 179, 194–95, 203, 217–18

Proportional-plus-integral control, 179, 191–92, 202

Proportional sensitivity, 190

Q

Quadratic formula, 74

R

Regula falsi method, 73, 78–80, 99

Resistive elements, 118, 141–42, 144, 168

Response functions, 11–12, 24

Reynolds number, 79

Rise time, 117, 138, 168

Robotics, 4, 9, 153

Root loci, 133

Root-locus method, 211, 220–30

Routh criterion, 212, 214

Routh-Hurwitz method, 211–20, 228–30

Runge-Kutta methods, 73, 88, 91–98, 100

S

Sampled-data systems, 235, 239–41, 245

Sampled signals, 239
Secant method, 73, 78, 98
Sensitivity, 1, 4, 6, 9
Sensors, 189
Servomechanisms, 3
Settling time, 117, 138, 157, 168, 220
Shift theorem, 46, 57–58, 240, 245
Signals, 2
Simulation, 6, 8, 167
Simultaneous differential equations, 31, 40
Sliding friction, 143
Socioeconomic systems, 167
Spring constant of proportionality, 141
Square-law behavior, 142
Stability, 4, 6, 9, 19, 39, 133, 180, 193, 211–35
State space modeling, 235, 243–45
State variables, 235, 245
Static friction, 142–43
Steady-state error, 157, 162, 179–210
Steady-state solution, 11, 14, 20, 38
Step size, 90
Successive approximations, method of, 73, 77–78, 98
Sum operator, 108
System analogies, 118, 140–56, 168
System analysis, 117–78
System components, 140–56
Superposition, principle of, 13, 24, 47, 107, 242

T

Taylor's series, 11, 26–27, 40, 75, 78, 88, 91
Temperature regulator, 3
Through variables, 117, 144, 168
Time constant, 117, 122, 168

Time lags, 1, 4, 9
Torque, 141
Torsional damper, 143
Torsional spring constant, 141
Torsional viscous damping coefficient, 142
Transducers, 189
Transfer functions, 47, 106–16, 181
Transformation calculus, 47
Transformation operator, 47, 68
Transient solution, 11, 14, 38
TUTSIM, 167
Two-position control, 179, 189, 202
Type number of a system, 179–82, 184–88, 202

U

Undamped behavior, 117, 131, 134, 168
Undamped natural frequency, 130, 156
Underdamping, 117, 131, 135, 168, 211

V

van der Waals equation of state for real gases, 83–84

W

Watt, James, 3
Wiener, Norbert, 4

Z

Zeroth-order systems, 117–119, 167
z-transform, 235, 239–41, 245